Modernismo, Modernity,
and the Development of Spanish American Literature

Modernismo
Modernity
and the Development of Spanish American LITERATURE

Cathy L. Jrade

UNIVERSITY OF TEXAS PRESS, AUSTIN

Excerpt from "The United Fruit Company" by Pablo Neruda (in *Canto general*, translated by Jack Schmitt, © 1991 Fundación Pablo Neruda, Regents of the University of California) used with permission.

"Poem II" of *Versos sencillos* (in *José Martí, Major Poems: A Bilingual Edition*) reprinted courtesy of Holmes and Meier Publishers, Inc.

Printed in the United States of America

First edition, 1998

∞ The paper used in this publication meets the minimum requirements of American National Standard for Information Sciences—Permanence of Paper for Printed Library Materials, ANSI Z39.48-1984.

LIBRARY OF CONGRESS

CATALOGING-IN-PUBLICATION DATA

Jrade, Cathy Login, 1949–
Modernismo, modernity, and the development of Spanish American literature / by Cathy L. Jrade. — 1st ed.
p. cm.
Includes bibliographical references and index.
ISBN 0-292-74049-2 (cl. : alk. paper). — ISBN 0-292-74045-X (pbk. : alk. paper)
1. Spanish American literature—19th century—History and criticism. 2. Modernism (Literature)—Latin America. I. Title.
PQ7081.J73 1998
860.9′112′098—dc21 98-5890

Texas Pan American Series

for Rachel

Contents

Preface

As is often the case, the central ideas for this book grew out of previous work. My *Rubén Darío and the Romantic Search for Unity,* also published by the University of Texas Press, focused on the head of the *modernista* movement and his poetic quest to fill the spiritual void of modern life. In the chapter on *modernista* poetry published in the three-volume *Cambridge History of Latin American Literature,* I began to explore other fundamental features of the *modernista* project. I showed that politics was an essential element of the movement and that this political vision was intimately tied to *modernismo*'s epistemological assumptions. As I developed this premise, I came to understand that *modernismo* was, for Spanish America, just the first in a series of complex and continuing confrontations with modernity. I therefore set out to place *modernismo* in a broader temporal framework, clarifying the way *modernismo* affected the works of those authors that came after.

Though the presentation of these ideas forms the core of the book, I did not want the beauty and power of *modernista* texts to be lost from sight. I have sought to maintain in the forefront the verbal grace, majesty, and force of the works while attempting to put finally to rest—through careful analysis and presentation—the persistent but erroneous belief that *modernismo* is a movement more concerned with form than content. This misconception has grossly shortchanged the far-reaching significance of the *modernista* project.

While I have addressed concerns of interest to specialists, I have also aspired to open the intellectual wealth and energy of the movement to those interested in other areas of Hispanic and world literature. To achieve this goal, I have translated all titles and texts into English, relying only occasionally on published translations. I have laid out the translations of poems in paragraph form to indicate that, in trying to remain as faithful as possible to the sense of the original, I have not attempted to reproduce the poetic structure. The slashes within these prose translations are meant to help readers who are unfamiliar with Spanish locate themselves in the original. Even here, however, English word order occasionally demands radical departures from the Spanish. So as not to clutter the page excessively, title translations appear only the first time that they are cited.

I am indebted to those who have taken the time to read the manuscript and offer suggestions. My heartfelt thanks go out to Roberto González Echevarría of Yale University for his insights, comments, and longstanding friendship. I am also thankful to Ana Eire of Stetson University for our ongoing exchange of ideas and for her assistance in capturing the nuances of some Spanish titles in English. To my colleagues and graduate students at Vanderbilt, I am grateful for a stimulating and productive environment. I thank my parents, mother-in-law, family, and friends for their quiet support and affection. For Ramón, no words of appreciation can possibly be enough. He has shared with me twenty-five years of living. In the process, he has shown me the strengths of the sociological way of thinking. He also encouraged me, throughout the writing of this work, to clarify difficult ideas and complex sentences. His loyalty and love are uncommon. Together we have raised our daughter, Rachel, to whom this book is dedicated and for whom a lifetime of love is offered in every word.

Finally, I wish to express my appreciation to Vanderbilt University and the University Research Council for the sabbatical leaves and various grants that have made it possible for me to research and write this book.

Modernismo, Modernity,
and the Development of Spanish American Literature

ONE

Spanish America's Ongoing Response to Modernity

In 1888 Rubén Darío chose the term *modernismo* to designate the shared orientations of Spanish American authors writing toward the end of the nineteenth century.[1] In choosing this label, Darío, head and intellectual center of gravity of the movement, was acknowledging an essential factor that for the most part has been overlooked by critics. He affirmed that what he and his fellow writers were attempting to do was to establish a mode of discourse commensurate to the new era that Spanish America had entered. The term chosen underscores the conviction held by those who adopted it as a banner of distinction and pride that this movement was born of and within the context of modernity.

The repercussions of this simple statement are multiple, but of utmost significance is its ability to clarify what has become a point of overheated debate among critics. Many who set out to specify what the modern—and now the postmodern—means for Spanish America feel compelled to point out that this region of the world and its social structures have not yet experienced "modernity" in Max Weber's sense of the term, referring to the increasing "rationalization" of life. Others concede that contemporary Spanish American culture and conditions are a product of uneven inroads of modernization.[2] Regardless of the extent to which Spanish American countries have diverged from the Anglo-European route to development or to which they continue to exhibit a "Garciamarquesian" fusion of premodern, modern, and postmodern influences, the Spanish American writers of the end of the nineteenth century—most of whom lived in the urban capitals of their countries and/or traveled extensively in Europe—believed that they were confronting, in a noble struggle, the most acute issues of modern life.

This basic point has failed to garner the attention it deserves.[3] Consequently, critics of Spanish American literature have overlooked important features of the artistic and cultural trends that crisscross the Atlantic starting at the end of the last century. My book will, accordingly, explore the nature of the connection between *modernismo* and modernity, thereby modifying the way in which the movement is perceived. I will show that the cultural and political transformations brought about by modern life engendered literary responses which differed in crucial ways from those of previous movements. Moreover, I will demonstrate that *modernismo,* because it is the first Spanish American movement to take up the challenge of modernity—in all its ramifications—ushered in fundamental shifts in the roles assigned to the poet, language, and literature. These changes have continued to influence artistic production to this day, ushering in further developments that give recent Spanish American literature equal claim to the much contested rubric of postmodernism.

Today it is well acknowledged that the distant beginnings of modernity are located within the Renaissance. The underlying philosophic break that took place—the one that shapes all successive thinking—pertains to the issue of the grounding of knowledge. The key characteristic of modernity arose as intellectuals cut themselves off from divine guarantees of knowledge. As Wlad Godzich, in his discussion of the relationship between modernity and postmodernity, points out, "[T]he problem that haunts all modern thinkers from Descartes, Locke, and Kant onward, is that of ensuring the reliability of knowledge (i.e., its legitimacy) and all forms of individual and collective action that rest on it" (114).

As distant as its philosophic origins may be, modernity's more specific characteristics took shape during the second half of the eighteenth century. These features are generally identified with scientific and technological progress, the Industrial Revolution, and the sweeping economic and social changes brought about by industrial capitalism.[4] Many contemporary theorists have focused on how this tangled interplay of philosophic and socioeconomic developments has defined the course of modernity and, now, postmodernity.[5] In his study on critical trends within the latter movement, John McGowan describes the situation that serves as the point of departure for modern times. "By the end of this period [1500–1800], the West has recognized, in the face of diversity and change, that it is thrown back upon itself to ground, legitimate, and make significant its own practices" (4). With the loss of belief in divinely given premises for human action, most intellectuals in Western societies began to experience a struggle for dominance among different sectors of society, within which different narratives and principles of legitimacy were established. Though the sense of loss, fragmentation, and alienation was widespread, the predominant conflict that arose was between materialistic and spiritual aspirations. Torn between the reigning faith in technology, progress, and empirical science and an enduring fascination with intangible realities and nonmaterialistic goals, writers and philosophers began to express their sense of being out of touch with themselves and the world around them.

The first major movement to focus on the issues of fragmentation and alienation was European romanticism. Its principal exponents sought to re-create the lost ethical totality of society by reappropriating premodern visions of life and language. As Godzich makes clear, the attraction to premodern visions reflects the quest for a culturally stable, as well as a politically just, anchoring of everyday life.

> Prior to modernity the relation [human beings] had to the world was taken to be one of knowledge, and this knowledge, on which individual and collective identities depended, was guaranteed by some divine instance or by some constitutive homology between humans and the world. Such a knowledge permitted humans to act, to build a world of human relations that increased the sum of knowledge—that is, their set of relations, their mode of being in the world. With the advent of modernity a change begins to take place in this economy: the old guarantees of knowledge cease to hold true and we are threatened with individual meaninglessness and collective tyranny, the latter understood as the arbitrary exercise of power. (127–128)

It is precisely the perception of the onset of this double threat—individual meaninglessness and possible political excess—that led Spanish American writers toward the end of the nineteenth century to follow in the literary footsteps of the European romantics. Just as romanticism challenged the hegemony of the scientific and economic in modern life, Spanish American *modernismo* protested the technological, materialistic, and ideological impact of positivism that swept Spanish America as it entered the world economy during the nineteenth century. Both the European and the Spanish American movements sought to provide an alternative view of existence that they claimed to be more inclusive. The *modernistas,* like the romantics before them, favored an alternative that was primarily "spiritualist," predicated on changes in consciousness and values. They proposed a worldview that imagined the universe as a system of correspondences, in which language is the universe's double capable of revealing profound truths regarding the order of the cosmos.

At the same time, however, *modernista* authors also manifested a "realist" tendency, seeking to establish more directly political and worldly images of change. They were acutely aware of their innovative position with regard to Spanish American literary history and literature's complex relationship with emerging national identities—the result of political consolidation following the wars of independence. They believed that they were creating for the first time a literary movement that would bring Spanish America out of its postcolonial isolation and its anachronistic backwardness into the modern present. *Modernismo* would establish a new mode of discourse for Spanish America, one that would reveal hidden realities as well as address issues of social and political consequence. In short, they sought to create a literary language with which to respond to their modern predicament, a language that, by being both spiritual and political, would make them equal to their European contemporaries.

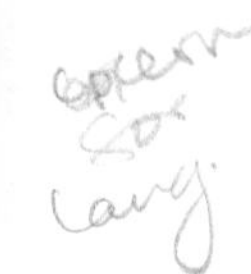

Thus there developed within *modernista* literature a concern for language that is dual in nature. One aspect is related to language as an instrument of vision and knowing, capable of revealing realities concealed by the inflexibility of scientific methods and the stultification of everyday concerns and values. The other is related to language as a tool of politics and power that plays an essential role in the formation of national cultures and identities. This second concern grows out of the first and derives its legitimacy from the movement's faith in the superior epistemological power of literature. In short, *modernismo* asserts its ability to comment on and even alter the positivistic, materialistic, and pragmatic course adopted by the Spanish American nations entering the modern age. Both aspects are derivative of the desire to use literature to influence the development of modernity. Both are grounded in the transformative capacity of art.

This phenomenon, which I will examine in the second chapter, had its antecedents at the onset of modernity in Spanish America. The end of the colonial era and the beginning of the period of independence brought changes that marked the transition to modern life. The debate that ensued between the old guard and the Young Turks anticipates some of the points made by the *modernistas.* But it is important to recognize the significant differences that begin with *modernismo.* The *modernistas* were the first writers to experience and appreciate the all-encompassing alteration in the fabric of life in Spanish America brought by modernity. The *modernistas* were the first to witness the tragic face of science as it robbed legitimacy from the religious, magical, and animistic worldviews that had ruled the daily lives of most Americans since before the arrival of Columbus. The *modernistas* were the first to define the poet as both visionary and outcast, at odds with the dominant social values while striving to reveal those aspects of reality hidden by habit and convention. No longer protected by a privileged and patronized position in society, the *modernistas* were the first to struggle with the newly commercialized social arrangements that were taking hold. The *modernistas* were the first to live the perhaps irreconcilable tension between the search for a spiritual community and a sense of national identity, on the one hand, and a longing to participate in the world arena, on the other. These are some of the factors that constitute the innovative *modernista* response to the modern world and that link *modernismo* to the movements that followed.

As late *modernista* tendencies turned into *la vanguardia* [the avant-garde], the artistic challenge to the emerging status quo remained strong, but the form it took changed. Insistence on art's autonomy became a dominant characteristic. Like Anglo-European modernists and avant-gardists writing at approximately the same time, the *vanguardistas* declared themselves answerable to nothing, pushed the limits of acceptable subjects and forms, and proclaimed a heady freedom from traditions, social environment, and reality itself. But if the Anglo-European modernists followed a strategy that asserted the independence of artistic creation and that made their confrontation with their social milieu less obvious, their Spanish American counterparts, prodded by a series of unique historical events, maintained, for the most part, a more directly political and self-critical stance. In this regard, they resemble more closely the writers of the European avant-garde.[6] All three groups, however, manifest traits that reveal their grounding in and exaggeration of trends that started in the nineteenth century in defiance of the dominant attributes of modern life. They pursue experimentation; they demonstrate a desire to *épater le bourgeois,* that is, a desire to shock middle-class society; and they believe that variations in perception and consciousness can drastically alter prevailing social forms.

Jean-François Lyotard's version of this trajectory of modern art centers on what he calls "the shattering of belief" and the discovery of the "'lack of reality' of reality" (*Postmodern Condition* 77). He relates this "lack of reality" to the aesthetics of the sublime through which he believes modern art finds its impetus and the avant-gardes find their axioms. For Lyotard, the sublime is that which cannot be represented adequately, whether that be the totality of the world or that which is infinitely great. "I shall call modern the art which devotes its 'little technical expertise' . . . to present the fact that the unpresentable exists. To make visible that there is something which can be conceived and which can neither be seen nor made visible . . ." (78). He goes on to suggest that the role of the avant-gardes is to humble and disqualify "reality" by exposing those techniques that make the viewer or reader believe in it.[7]

Within Spanish America this distrust of hegemonic discourse and its facile assumptions about reality appears in texts throughout the first half of the twentieth century, in challenges to traditional historical perspectives, conservative political positions, and simplistic positivistic epistemologies. Gradually, however, a shift occurs: the critique becomes all-encompassing, and the old belief that the distortive constraints of the dominant culture could be supplanted by a superior artistic perspective is replaced by the recognition that the critical distance previously desired cannot be achieved by artists who function within, and thereby acquiesce to, the ever more intrusive social constraints. Instead of anchoring the individual in the world, ensuring some form of stability, knowledge is seen as a coercive force. The artist in particular is aware of its structuring nature and, accordingly, its capacity to control and to dictate compliance. Accounts of the legitimation of knowledge are suspect and examined for inconsistencies and errors. The result is a progressive loss of faith in "master narratives," a loss of legitimation.

The earliest manifestations of these "postmodern" trends can be found in the writings of Jorge Luis Borges and develop—unevenly—throughout the rest of the twentieth century. While movements are never monolithic and older features tend to survive well into periods of innovation, part of the confusion surrounding postmodernism is the longevity of romantic and postromantic (modern) literary characteristics, which reappear constantly. This situation is further complicated by two additional factors. First, the artistic tendencies associated with postmodernism precede by at least two decades, and then accompany, the flurry of theoretical writings that have turned postmodern concerns into a cottage industry among critics, philosophers, intellectuals, and scholars. Second, "the postmodern" is a label that is now used to designate cultural trends that are linked to socioeconomic and political developments of late capitalism and postindustrial society with its virtual reality, electronic communications, and cyberspace. This latter use becomes particularly problematic for those concerned with Spanish America, for it raises questions about the degree to which Spanish America partakes of these developments or is victimized by them.[8]

Upon careful scrutiny, it becomes evident that postmodern artistic responses to society are both a continuation and a departure. They persist in seeking to undermine the taken-for-granted underpinnings of power and privilege, while going much further. Postmodern works tend to reject foundational philosophies and related totalizing beliefs, such as those embodied in humanism, rationalism, imperialism, and patriarchy. All of these ideologies imply an ordering of dominant and subordinate hierarchical divisions, which permits the often exploitative privileging of one aspect at the expense of another. Instead, postmodern works seek to undermine hegemony by offering their particular response to modern or postmodern life in the spaces between positions, that is, in the interstices between political, epistemological, and discursive stands. The hope for disruption of the hegemonic structures is seen in empowering those elements within society that have traditionally been suppressed. As a result, postmodern artistic production seeks to overcome distinctions such as those made between high and low art, between artist and critic, between prose and poetry, between signified and signifier. The often disconcerting results are the constant explosion of assumptions, the constant collision of modes of speech, the constant shuffling of possibilities evident in the works of the late twentieth century.

Lyotard's aesthetics of the sublime adds yet another dimension to this discussion. He writes:

> Here, then, lies the difference: modern aesthetics is an aesthetic of the sublime, though a nostalgic one. It allows the unpresentable to be put forward only as the missing contents; but the form, because of its recognizable consistency, continues to offer to the reader or viewer matter for solace and pleasure. Yet these sentiments do not constitute the real sublime sentiment, which is in an intrinsic combination of pleasure and pain: the pleasure that reason should exceed all presentation, the pain that imagination or sensibility should not be equal to the concept.
>
> The postmodern would be that which, in the modern, puts forward the unpresentable in presentation itself; that which denies itself the solace of good forms, the

consensus of a taste which would make it possible to share collectively the nostalgia for the unattainable; that which searches for new presentations, not in order to enjoy them but in order to impart a stronger sense of the unpresentable. A postmodern artist or writer is in the position of a philosopher: the text he writes, the work he produces are not in principle governed by preestablished rules, and they cannot be judged according to a determining judgment, by applying familiar categories to the text or to the work. Those rules and categories are what the work of art itself is looking for. (*Postmodern Condition* 81)

Both the modern and the postmodern strive to assert what lies beyond the realm of everyday reality, but the postmodern is aware that any recourse to established modes of presentation will forestall the possibility of achieving this goal.

The trajectory that has just been traced begins in Spanish America with *modernismo*. Looking at Spanish American literature as a continuing response to modernity enhances our understanding not only of *modernismo* but also of crucial aspects of succeeding trends and developments. Specifically, through a sustained examination of the entire movement, this study will show how *modernismo* represents a key, perhaps *the* key, episode in Spanish America's many and continuing literary responses to its incorporation into the world economy and its introduction to modern life. Through detailed analysis of major works by the most prominent *modernistas*, this study will explore the impact of the social and political dislocations that were brought about by the rapid modernization of Spanish America during the second half of the nineteenth century. These realignments not only radically altered the traditional, feudal-like social arrangements that had characterized Spanish America for centuries but also laid the foundation for life and thought in the twentieth century.

An examination of this common ground will highlight how and why *modernismo*'s two central tendencies pertaining to epistemology and politics surface and submerge at different points and with different authors. While, for the most part, the search for a language capable of revealing realities unobserved by those caught up in the activities of modern life appears to take priority and lock *modernismo* within a perspective that is either aesthetic or spiritual, this very concern immediately extends itself into the realm of sociopolitical commentary. The ideal language reveals truths that have the power to alter the ignorant assumptions of the uninitiated. The resulting knowledge provides the basis for artistic, spiritual, moral, and political decisions.

Chapter 2 focuses on the *modernista* movement in general and attempts to place it in a broad context. It examines the interplay of the artistic, philosophic, social, and political trends that influenced the way *modernista* writers came to formulate their predicament, their mission, and their goals. It revisits *modernismo*'s immediate literary antecedents and surveys early perceptions of the movement. Most importantly, however, it refocuses the analysis of these developments, redirecting attention to those aspects that have been previously overlooked or underappreciated. This overview shows how, in *modernismo*'s response to modernity, epistemology and politics become interwoven as a foundational concern of the movement, one that informs succeeding literary endeavors.

Chapter 3 examines the early years of the movement and deals with the four most important *modernistas* of this period: Manuel Gutiérrez Nájera (Mexico, 1859–1895), José Martí (Cuba, 1853–1895), Julián del Casal (Cuba, 1863–1893), and José Asunción Silva (Colombia, 1865–1896). The chapter highlights specific texts by these early *modernistas* and shows how the tendencies outlined in chapter 2 take shape during this period. By including both prose and poetry, it underscores the generalizability of the conclusions reached.

Just two years younger than the youngest of this first group, Rubén Darío (Nicaragua, 1867–1916) was both their contemporary and their successor, for, by 1896, all four writers studied in chapter 3 were dead. Because of this unique turn of events as well as his extraordinary talents and vision, Darío became the leading *modernista* and his work came to represent the epitome of *modernista* art, embodying the entire breadth of the movement. Chapter 4, therefore, looks at his entire production, examining the personal and generational issues he brought to bear on the course of the movement with which he became identified. It shows that Darío's confrontation with modernity—with its undermining of aesthetic values, of spiritual concerns, and of the privileged place in society for the poet—is evident throughout his career and in nearly everything that he wrote.

Chapter 5 deals with the diverse group of poets who, in tandem with Darío, contributed to the development of the movement. They are Enrique González Martínez (Mexico, 1871–1952), Amado Nervo (Mexico, 1870–1919), Ricardo Jaimes Freyre (Bolivia, 1868–1933), Guillermo Valencia (Colombia, 1873–1943), José María Eguren (Peru, 1874–1942), and José Santos Chocano (Peru, 1875–1934). With careful consideration of specific texts, the chapter focuses on the authors' continued commitment to the epistemological and political concerns that shape *modernismo*. At the same time, the chapter endeavors to reveal how these writers sought to adjust the nature of their discourse in accord with their constantly evolving artistic and social contexts.

While chapter 5 examines those authors whose works demonstrate *modernismo*'s evolution within a framework of continuity, chapter 6 centers on three poets whose works brought *modernismo* to the threshold of the avant-garde. The poetry of Leopoldo Lugones (Argentina, 1874–1938), Julio Herrera y Reissig (Uruguay, 1875–1910), and Delmira Agustini (Uruguay, 1886–1914) reveals how *modernista* tendencies were altered by changing philosophic perspectives, sociopolitical pressures, and an astutely critical view of poetic language. These three authors, all of whom are from the southern cone of Spanish America, reinvigorated and radicalized the *modernista* project while remaining true to the movement's primary concerns.

These concerns with offering an informed and enlightened alternative to the dominant and coercive vision of life imposed by the forces of modernity run throughout *modernista* works. However, as *modernista* images, style, and tone became associated with affectation and effete elegance, it became progressively more difficult for readers to perceive this critical response to modernity. By combining broad overviews and detailed analyses, I have aspired to reclaim the power of the visionary stance taken by these creative intellectuals. I have also tried to indicate, now with the gift of hindsight, the many ways *modernismo* anticipates features, themes, and beliefs that continue and develop in later literary movements. Contemporary Spanish American writers have acknowledged their ties to *modernismo.* The final chapter therefore briefly comments on *modernismo*'s reflection in more recent literary production.

As already indicated, Lyotard's opinions on the centrality of the sublime offers a way of envisioning the type of developmental relationship between modernity and postmodernity that I am proposing. Even before him, however, Carlos Fuentes refers to this significant epistemological phenomenon as the return to the poetic roots of literature by novelists of the twentieth century. Though *modernista* works are not mentioned in this context in *La nueva novela hispanoamericana,* my study seeks to facilitate the recognition of such connections. Fuentes writes:

> In order to imagine the path that the novel will take in a world that we still cannot name, it will be necessary to think first about writers like William Faulkner, Malcolm Lowry, Hermann Broch, and William Golding. All of them returned to the poetic roots of literature, and through language and structure, and no longer thanks

to complex plots and sociology, they created a representative convention of reality that aspires to be totalizing. It invents a second reality, a parallel reality, ultimately a space for *the real,* through a myth in which the hidden half of life as well as meaning and unity of diffuse time can be recognized, but is no less true because of its mythic nature. (19)[9]

This shift to "the poetic," to "the sublime," underscores a fundamental dissatisfaction with the dominant Western perspective that has emphasized materialism, rationalism, and pragmatism since the beginning of modern times. This disillusionment with modern values and ways of knowing first appeared in Europe with romanticism and with *modernismo* in Spanish America.[10] Starting at the beginning of the nineteenth century, the struggle between appearance and reality, between words and things, followed a distinctive course under the influence of factors that have come to define modern life. These, in turn, generated literary responses throughout the West. In Spanish America the story begins with *modernismo.*

TWO

Modernismo: Knowledge as Power

ATTEMPTS to analyze Spanish American *modernismo* have now entered their second century. Virtually from its inception, literary observers and critics have struggled to pinpoint its distinctive nature and to detail its primary characteristics. Though Rubén Darío defined his—and, by implication, *modernista*—aesthetics as "acrática," that is, opposed to all authority, and even though this feature continues to appear on the lists of fundamental characteristics of *modernista* verse, *modernismo* manifests an essential unity that stems from its origin in a shared literary, philosophic, and social context.[1] With its faith in the poet and poetry, *modernismo* proposed a profound response to the crisis of beliefs that surfaced among the philosophers and artists of Spanish America toward the end of the nineteenth century, a crisis similar to the one that had dominated intellectual circles throughout the West since the onset of modernity. With its attention to language as a cultural resource capable of capturing, defining, and elucidating issues of national identity and autonomy, it addressed the quandaries faced by postcolonial Spanish America.

Modernismo began in Spanish America in the late 1870s and lasted into the second decade of the twentieth century. It was the inventive nature of its language together with the adaptation of diverse literary sources, especially the strong influence of French poetry, that first attracted attention.

In his famous letter to Darío dated 22 October 1888 and written in response to receiving a personally dedicated copy of *Azul . . .* [*Blue . . .*], Spanish critic and novelist Juan Valera defined Darío's originality and technical perfection in terms of his "galicismo mental" ["mental Gallicism"] (25). Though Valera's judgments are colored by nationalistic pride and ideological conservatism, one can find in his letter to Darío those features of the movement that have repeatedly received detailed critical attention during the more than one hundred years since it was written. From the outset, major studies sought to pinpoint French models for *modernista* works, with the earliest criticism focusing primarily on the formal aspects of this influence.[2] Scholars emphasized changes in meter and verse form, the introduction of new rhythm and rhyme schemes, and symbolic and thematic similarities with earlier French texts. Others chose to distinguish between the perfection of form and devotion to beauty attributed to the influence of Parnassian verse from the musical evocation and dreamlike suggestion encouraged by symbolist poetry.[3]

This line of inquiry was pursued by Pedro Salinas ("El problema," orig. pub. 1941) and Guillermo Díaz-Plaja, both of whom continued to characterize *modernismo* as limited in scope, that is, primarily concerned with aestheticism, the search for beauty, and the renovation of poetic form. Even Max Henríquez Ureña's comprehensive, insightful, and influential *Breve historia del modernismo* reflects this general orientation. In it he holds that the central concern of *modernista* authors was to break with "the excesses of romanticism" and "the narrow criterion of pseudoclassical rhetoric" (13–14). He believed that, for this reason, they opted to transpose French innovations into a Spanish key. Repeated and detailed references to Leconte de Lisle, José María de Heredia, Alfred de Musset, Victor Hugo, Paul Verlaine, Catulle Mendès, Stéphane Mallarmé, Maurice de Guérin, Théophile Gautier, Pierre Loti, among many others, reaffirm this focus.

The search for "sources" and influences was not, however, limited to France. As Valera noted with regard to *Azul . . .* , there is a strong and wide-ranging cosmopolitan spirit that runs through *modernista* works. All of European and even Middle Eastern and oriental art and culture captured the poetic imagination of *modernista* authors from time to time.[4] Occasionally the influence was direct, more often it came by way of Paris, filtered through Parisian imaginations and interpretations.[5] While this cosmopolitanism has been identified with escapism, a rejection of the stifling restrictions of Spanish poetics and culture, and flight from the immediate Spanish American reality, it is actually a manifestation of a complex and profound search, a search that led *modernista* writers to embrace diverse aspects of high culture from all corners of the globe with a heady enthusiasm in the expectation of achieving—in apparent contradiction—a sense of identity that is clearly Spanish American.

Aware of their extraordinary place in Spanish American history, *modernista* writers broke with Spanish models which they understood to be both grandiose and inflexible. They questioned the ability of Spanish, with its conservative rules and archaic associations, to express either their new sense of self or their perceptions of the changing social scene. They turned their eyes instead toward the rest of Europe in order to help them define the present and, through the present, the future. This attitude is evident in Darío's selection in 1888 of the term *modernismo* to designate the tendencies of Spanish American poets (see Henríquez Ureña 158–172). This choice underscores the *modernistas'* desire to be modern, that is, to become contemporaneous with all of Europe but most especially with its intellectual center, the city of Paris. The poets sought to leave behind—either through their travels or their imagination—an anachronistic, local reality in order to establish for themselves a modern mode of discourse in which they could speak for the first time with their own voice and with an unclouded, critical vision of Spanish America.

As a result, "escapist literature" almost immediately became, as noted by Octavio Paz in "Literatura de fundación" (11–19), a literature of exploration and return. *Modernista* writers turned their attention from the most up-to-date European trends toward home and resurrected, through flights of fancy as much as through historical fact, a Spanish American past that included ancient civilizations, indigenous peoples, and a Spanish American consciousness. By acknowledging and embracing indigenous cultures as part of their desire to formalize and to found a modern Spanish American discourse, the *modernistas* not only asserted their intrinsic difference from their Spanish, Anglo-European, and North American contemporaries but also affirmed what they considered to be an ancient advantage over more "recent" European traditions. On an aesthetic plane, recourse to various aspects of indigenous culture provided an unconventional source of artistic inspiration, which added a uniquely Spanish American dimension to their unfettered pursuit of beauty throughout the centuries and across all borders. At the same time, the *modernistas'* real and immediate ties to premodern modes of perception and belief offered a reservoir of responses to the modern world that they had entered. While this investigation of native life tends to take a backseat to the *modernista* fascination with European traditions, it emerges shortly thereafter as central to the *novela de la tierra* and continues to appear throughout the literature of the twentieth century.[6]

Modernista writers also grappled with the various modes of discourse that were vying for dominance toward the end of the nineteenth century. They considered the poetic and prosaic, the religious and scientific, in their attempt to find their own voice. Aníbal González, in his groundbreaking study *La crónica modernista hispanoamericana,* explains the complex relationship between philology and *modernismo.* He begins with Renan's simple but powerful premise that "[t]he modern spirit, that is to say, rationalism, criticism, liberalism, was founded the day that philology was founded. *The founders of the modern spirit were philologists*" (*L'Avenir de la science,* qtd. in González 5; emphasis in original).[7] He goes on to analyze how, in philology, *modernista* literature, particularly the prose, finds both a model and an antagonist. Philology provides a way of knowing based on language's ability to recover the past and reveal what is hidden. Yet philology's confrontation with the opacity of language provides a space that only literature can fill (12, 14, 16, 54–56). In other words, literature grants itself a privileged status over more "scientific" ways of knowing, maximizing in this way its ability to counter the hegemony of the materialistic and economic in modern life. This position was part of a more general "spiritualist" response by Spanish American intellectuals, which they hoped would be more inclusive. They understood their goals to be profound and far-reaching and their efforts to be simultaneously philosophic, aesthetic, and political. The balance among these three aspects tilted toward political concerns only as the socioeconomic pressures that gave rise to *modernismo* exploded in crisis in 1898 with the Spanish American War and later, in 1903, with the carving out of the Panamanian state as a result of United States intervention.[8]

Most critics who have discussed the "political" nature of *modernista* discourse focus on this later period and its more overtly nationalistic agenda, but they tend to disregard its more subtle origins in the writings before the turn of the century. The "politics" of *modernista* writing grows out of its assertion that it holds the key to a profound understanding of realities overlooked or ignored by the ever increasing emphasis upon pragmatism, materialism, empiricism, and "progress." These realities, it was believed, form the basis of all action appropriate to orderly, harmonious, and moral behavior. The *modernista* attempt to find a language with which to communicate its epistemological insights and to present its implicit ethical stance is neither unilinear nor consistent, yet at all times it demands a committed engagement with the most crucial questions of the day. *Modernista* art reflects a confidence that it can, with its superior understanding of the workings of the world, have a constructive and beneficial role in confronting the radical changes brought about by integration into the modern world economy and in formulating alternative paths of national development. This perspective is profoundly political in nature.

The change from what was erroneously believed to be an apolitical perspective, as in the poetry of Manuel Gutiérrez Nájera and Darío's early works, to more assertively Spanish American concerns, as with key poems in Darío's *Cantos de vida y esperanza* [*Songs of Life and Hope*] and *Alma América: Poemas indo-españoles* [*Soul America: Indo-Spanish Poems*], by José Santos Chocano, led early critics to posit two distinct stages in *modernismo.* The "second generation" of *modernistas* was considered more focused on achieving an artistic expression that would be genuinely American. This misleading division overlooks the fact that both "generations" dealt with the issues of Spanish America's place in the modern world and the creation of a language and vision appropriate to that place. The recognition of the centrality of these issues to the entire movement not only reveals *modernismo*'s essential unity but also explains why the patterns set within the *modernista* movement reappear in literary works throughout the following century.

The socioeconomic conditions that most directly affected the development of *modernismo* of course vary from country to country. There were, however, certain key factors behind these variations, the most important of which were the strengthening of the national state and the accelerated integration of local economies into world markets. For the most part, the last decades of the nineteenth century saw a consolidation of state power that brought about a new degree of political stability—despite the periodic resurgence of caudillismo and anarchistic tendencies. At the same time, economic reorganization and growth brought prosperity and affluence to the upper classes. In urban centers, wealth and international trade encouraged a perceptible Europeanization of life. As Roberto González Echevarría has expressed it, in exchange for its raw materials, Spanish America received culture, primarily in the form of manufactured products ("Modernidad" 159). The turn-of-the-century flood of luxury items filled the homes of the old landed aristocracy, the nouveaux riches, and the aspiring bourgeoisie. It also created an image of life that left a lasting impression on the poetic imagination of the writers of the time, an image that evoked the sense of well-being, ease, and fashionable excess characteristic of the Parisian Belle Epoque—that is, of life in Paris during the three decades beginning with the 1880s.[9]

Members of the ruling class allied themselves with foreign financiers and investors, and their primary ambition became the accumulation of capital at the expense of more traditional goals and obligations. The political philosophy of the day was the positivism of Comte and later that of Spencer, both of which provided the foundations for a type of social Darwinism. Comte had developed a philosophical system that rejected metaphysics and relied exclusively on the positive sciences. His final aim was to reform society so that all men could live in harmony and comfort. During the peace that followed the political consolidation of the 1860s, positivism became the philosophy of order, promoting progress through empirical science and free enterprise. Society in Spanish America was to be organized upon a more rational basis than ever before. Scientists were believed to be the bearers of a demonstrable truth and the trustees of an infinitely superior future. Whatever evils of "modern life" arose were accepted as a necessary by-product of national development. In reality, however, the latent function of positivism was to provide the ruling classes with a new vocabulary to legitimate injustice; liberal ideology was gradually replaced by the belief in struggle for existence and the survival of the fittest. Inequalities were no longer explained by race, inheritance, or religion, but as the unfortunate but requisite consequences of progress. The Mexican dictator Porfirio Díaz and his circle of "científicos," the oligarchy of the Argentine landowners, and the Chilean nitrate barons epitomized the power holders who relied upon positivist ideology to justify their privileges.[10]

Positivism generated in most *modernistas* a strongly ambivalent attitude. They maintained a respect for science, its breakthroughs, and its contributions to progress; they rejected it, however, as the ultimate measure of all things. Despite the promises made, it became clear that, far from becoming more understandable, life appeared more enigmatic—the great inventions and discoveries had not provided answers to the fundamental questions of existence. If anything, Spanish America's growing prosperity and its increasing involvement with the industrial capitals of the world brought about social dislocations that heightened the sense of crisis among its writers. The new social context in which *modernista* art developed was marked by two essential elements: the disappearance of the old aristocracy, which had provided patronage of poetic production, and the transformation of all products of human enterprise—including art—into merchandise (Pérus 56, 66, 81). In this situation, poets had to earn their living producing a marketable commodity. Many supported themselves as journalists at the same time that they sought, through their well-crafted poetry, to assert themselves in a world where the items of highest esteem were luxurious, opulent, and usually imported.[11] Some, like Julián del Casal, became marginalized, creating a bohemian response to the vulgarity and utilitarianism of bourgeois society. Others, like Darío in his famous "El rey burgués" ["The Bourgeois King"], scorned the materialism, mediocre conformity, and aesthetic insensitivity of the growing middle class. Still others, like Martí, put their faith in the superior individual, "el hombre magno," who could see beyond the pressures of rapid urbanization and commercialization.

With these conditions, modernity, as it is understood in Western culture, arrived in Spanish America—or, at the very least, to its great, cosmopolitan urban centers. The ideological adjustments necessitated by the far-reaching alterations in the fabric of life that accompanied modernity generated a literary response, which was not unlike previous responses. As Octavio Paz notes in *Los hijos del limo* [*Children of the Mire*], modern poetry has always represented a reaction against the modern era and its various manifestations, whether they be the Enlightenment, critical reason, liberalism, positivism, or Marxism (*Los hijos* 10).

Matei Calinescu clarifies the nature of this antagonistic relationship between modernity and the literary reactions that it generates: ". . . at some point during the first half of the nineteenth century an irreversible split occurred between modernity as a stage in the history of Western civilization . . . and modernity as an aesthetic concept" (41). On the one hand, what Calinescu calls the "bourgeois idea of modernity" picked up and continued the tradition dominant within earlier periods in the history of the modern idea, emphasizing the doctrine of progress, the cult of reason, the ideal of freedom, and confidence in the beneficial possibilities of science and technology. All these features were reinforced by an ever stronger capitalist orientation toward pragmatism and by the cult of action and success held sacred by the middle class. On the other hand, "modernity as an aesthetic concept," which begins with the romantics and continues through the avant-gardes, manifests radical antibourgeois attitudes. This other modernity turned against the middle-class scale of values and expressed its disenchantment in many different ways, ranging from offensive effrontery to aristocratic self-exile.[12] Calinescu claims, in short, that "what defines cultural modernity is its outright rejection of bourgeois modernity" (42). Consequently, cultural modernity actually operates as a type of "anti-modernity."[13]

The first Spanish American variation on the tug of war between the two "modernities" is *modernismo.* Spanish American *modernismo* offers a response to the spiritual and aesthetic vacuum created by the positivist critique of religion and metaphysics as well as by the positivist support of materialistic, bourgeois values. It is therefore not surprising that, as the *modernistas* formulated their reaction to modernity and sought to deal with their feelings of alienation and anguish, they discovered appealing paradigms in the European literature that they had rushed to read in their attempt to create a poetic language consonant with modern times. They found appropriate models in English and German romanticism, French Parnassianism, and symbolism, for these literary movements too had been reactions to the spiritual upheavals generated by modern life. Broadly stated, European romantic and postromantic literary production influenced the way that *modernistas* came to formulate the general crisis of beliefs that has dominated Western culture since the Enlightenment and, perhaps more importantly, the poetic solutions that these writers opted to follow.[14]

Like the English and German romantics and the French symbolists, the Spanish American *modernistas* traced the anxiety of their age to fragmentation: individuals were out of touch with themselves, with their fellow humans, and with nature. Neither traditional religious beliefs, vitiated by liberal thought, nor the dry intellectualization of positivism provided satisfactory answers. They longed for a sense of wholeness, for innocence, for the paradise from which they had been exiled by the positivist and bourgeois emphasis on utility, materialism, and progress. The hope for amelioration resided in integration and the resolution of conflict. The design that the romantics elaborated for possible recovery and that was later adapted by the symbolists and the *modernistas* drew on ancient images of recuperation and unity.[15] Similarly, the language with which they confront the limitations of modern rationalism is rooted in the ancient tradition of analogy, that vision of the universe as a system of correspondences in which language is the universe's double.[16]

M. H. Abrams has shown that the basis of romantic thought lies in "highly elaborated and sophisticated variations upon the Neoplatonic paradigm of a primal unity and goodness, an emanation into multiplicity which is *ipso facto* a lapse into evil and suffering, and a return to unity and goodness" (*Natural Supernaturalism* 169). The myths and images that constitute this paradigm came to the romantics from all quarters of Western civilization. One of the most fundamental myths depicts primordial man as a cosmic androgyne who has fallen into evil and multiplicity yet retains the capacity for recovering his lost integrity. This myth has its roots in

Plato's *Symposium,* in Gnosticism, and in the Orphic and other mysteries; it resurfaces during the Renaissance and constitutes a central component of esoteric tradition.

As I have shown in *Rubén Darío and the Romantic Search for Unity,* many of these images of harmony and unity, which appear as alternatives to the disruptive forces of "bourgeois modernity," entered literary circles through occultism. The "occult sciences," as they came to be known, explored the "occult" mysteries of life that were ignored by conventional "science" while using the language of science, thereby giving a modern basis of legitimacy to their spiritual quests. Darío discerned the power of this phenomenon. When he wrote about the revival of occultist creeds in Europe for *La Nación* in 1895, he described their overall goal in the following manner:

> The science of the occult, which before belonged to the initiated, to the experts, is reborn today with new investigations by wise individuals and by special societies. The official science of Westerners has still not been able to accept certain extraordinary—but not out of the realm of the natural in its broadest sense—manifestations like those demonstrated by Crookes and Madame Blavatsky. But fervent followers hope that, with the successive perfecting of Humanity, there will come a time when the ancient *Scientia occulta, Scientia occultati, Scientia occultans,* will no longer be arcane. The day will come when Science and Religion, fused, will have man achieve knowledge of the Science of Life. (qtd. in Anderson Imbert, *La originalidad* 203)[17]

In addition to seeking a reconciliation between modern science and ancient wisdom, this passage underscores several key elements of turn-of-the-century society. The general malaise of the period—both in Europe and in Spanish America—is identified with a sense of loss of access to age-old knowledge. What Darío calls "the official science of Westerners" is faulted for considering only the physical aspects of life and for disregarding the ultimate questions of existence.

This passage also suggests that intellectuals, writers, and concerned individuals turned to the esoteric tradition that had survived throughout the centuries in hermetic and occultist sects in search of appealing paradigms with which they could confront the alienation of modern life and fill the void left by the collapse of orthodox belief systems. This tradition offered a vision of integration that satisfied the longings of those haunted by the ill effects of modernization.[18] The timelessness of ancient myths and images of unity provided an alternative to the prevailing sense of change, discontinuity, and fragmentation. Faith in a living, orderly universe supplied a response to the commodification of existence in which all elements of life are turned into objects to be bought and sold. Through these images, literature assigned itself the supreme task of decipherer of truths. Though these truths were primarily religious in nature, they clearly called into question choices about fundamental values and the use of material resources, establishing a dual vision that was simultaneously philosophic and practical.

As a result of their enormous appeal, numerous occultist sects enjoyed an extraordinary revival in Europe during the second half of the nineteenth century. They included the Theosophical Society, founded by Madame Blavatsky and Colonel Olcott; the Rosicrucians, headed by Sâr Péladan; and the Independent Group of Esoteric Studies, directed by Gérard Encausse, "Papus." They embraced a wide range of belief systems not always in perfect ideological harmony with one another. There were references to cabalism, astrology, magnetism, hypnotism, gnosticism, freemasonry, alchemy, and several oriental religions. Literary circles were filled with believers and proselytizers, and romantic and symbolist writings were permeated with cabalistic, hermetic, theosophical, and Eastern thought.[19] *Modernista* literature, in turn, reflects both the direct and indirect influence of these occult sources.[20]

Harmony as a philosophical ideal associated with divine perfection forms the basis of the ancient cosmology that structured most of the occultist beliefs that made their way into romantic and symbolist literature. The entire universe is held to be one harmonious and orderly extension of God, whose soul permeates all and is identical with the great soul of the world. Since both the individual and the universe are made in the image of God, each being is a microcosm that should strive to implant in his or her soul the harmony seen in the macrocosm. It is recognized, however, that there are certain superior individuals who are more conscious than others of the divine element within them and, consequently, more able to recognize the transcendent order of the world around them. Romantic and symbolist literary theory encouraged the identification of these special individuals with poets. Esoteric beliefs linked this superior status to a highly evolved soul, one that had become perfected through numerous incarnations. While not every *modernista* held fast to the idea of transmigration of souls, it is a concept that appears throughout the literature either playfully, as a dimension of *modernista* syncretism, or seriously, as an alternative to Christian salvation.

Occultism supplied another alternative to the strongly restrictive views of Christianity in its views of sexuality. Whether sexuality became important as a personal and immediate response to the emptiness of life or as a defiant challenge to the conservatism of middle-class morality—a defiance similar to dandyism in other aspects of life—esoteric tradition provided a way of looking at sex that made it possible for erotic longings to be incorporated in and to become an essential element of the *modernista* cosmology. As Paz recognized with regard to the romantics, the exaltation of the natural order of things, especially sexuality, is simultaneously a moral and political critique of civilization and an affirmation of a time before history (*Los hijos* 56–60). For virtually all the *modernista* writers, erotic passion is the most easily identifiable aspect of nature that has been inhibited or destroyed by the social order. By reclaiming the importance of sexuality, these authors hoped to perceive and understand the natural order of things hidden behind social conventions.[21] In other words, eroticism appears as a counterdiscourse to social expectations, one that reveals the failings of routine and habit.

For European poets of the second half of the nineteenth century and for the Spanish American *modernistas,* the occult sciences provided a conceptual framework for an openly erotic stance. By affirming the sexual nature of the godhead and by appropriating the esoteric myth of primordial man as a cosmic androgyne, sexual love became a means of approximating the androgynous state of the primal man. Since his fall into evil is identified with his entrance into the material and bisexual world, a return to the union of male and female became a means of perceiving the prelapsarian bliss of unity as well as of intuiting the divine state. The ideal female through whom the poet hopes to achieve a newly unfettered vision of the world is often linked with poetic language; and creation—whether literary, personal, or cosmic—is conceived of as sexual.

Darío's "Palabras liminares" ["Liminal Words"] from *Prosas profanas* [*Profane Hymns*] provides a revealing example of this phenomenon. When Darío in his introductory statement refers to the content of the collection and its title, he directs attention toward sexual passion—a sexual passion that fuses with art, poetic creation, music, and religion. He writes: "I have said, in the pink Mass of my youth, my antiphons, my sequences, my profane proses" (180).[22] Darío plays with the medieval allusions, breaks expectations regarding the genre in question, and equates divine love and religious devotion with sexual exploits. While pleasure is certainly at issue here, so is a great deal more. As Javier Herrero has pointed out, this blasphemous religiosity consists of replacing the Christian gospel with a new one in which the altar is presided over by Venus. Darío aspires to a mystical experience—radically different from those of Catholic mystics—that reveals the meaning of the universe, life, and art. His poetic renovation proposes a revolutionary change in values (40–43).

Darío's preoccupation with sexuality is intimately tied to his fascination with the constraints imposed on behavior, language, and vision by society. As a result, the sociocultural context of *modernismo* is never far from his mind. He begins "Palabras liminares" with regret over the lack of enlightenment among the general and educated public. It is art that sets him apart, but art is not imitation; it is the reinterpretation and revitalization of habit and custom by each artist. He declares: "I do not have a literature that is 'mine'—as a masterly authority has declared—in order to mark the direction of others: my literature is *mine* in me; he who follows slavishly my footsteps will lose his personal treasure and, page or slave, will not be able to hide the stamp or uniform. One day Wagner told Augusta Holmes, his student: 'The first rule, do not imitate anyone, above all, not me.' A great saying" (*Poesía* 179).[23]

This desire to reach beyond the social and conceptual frameworks that adulterate existence was widespread. For English romantics such as William Wordsworth, Samuel Taylor Coleridge, and Thomas Carlyle, poets were the individuals whose vocation was to liberate the vision of readers from the bondage of habitual categories and social customs so that they could see the new, harmonious world that the poets had come to see. For Charles Baudelaire and Arthur Rimbaud and other French symbolists, poets were the ones who could perceive the harmonious order of the universe behind the chaotic appearance of everyday reality.[24] In their poetry, as in Baudelaire's influential sonnet "Correspondances" ["Correspondences"], the disordered material of the mundane world is rearranged into an artistic creation that reflects the "dark and profound unity," that is, the orderly soul of both the visionary and the supernatural.

The special language through which the macrocosm and microcosm reveal themselves to each other is the language of symbols, metaphors, and analogies. The mission of poetry is to rediscover this means of communication and to achieve a renewed unity of spirit. To this end, Baudelaire encouraged the free use of words and images, which are to be employed not according to their logical usage but rather in accord with universal analogy, that is, emphasizing the "correspondences" between the material world and spiritual realities as well as among the different human senses. This interrelationship among the senses, which occurs because of affective resonances that cannot be accounted for by logic, provides the theoretical support for synaesthesia. Synaesthesia is often identified as one of the most distinctive characteristics of both symbolist and *modernista* verse, helping to turn it into an "evocative magic."

Going beyond Baudelaire, Stéphane Mallarmé sought to increase the magical powers of poetry by separating the crude and immediate from the "essential" condition of words. The "essential" word does not function as an intermediary between two minds, but as an instrument of power capable of awakening the soul to its original innocence. When restored to its full efficacy, language, like music, evokes a pure, untarnished view of the universe. Music, because it is indefinite and innocent of reference to the external world, became the ideal of poetic creation. Paul Verlaine had written, in his widely influential "Art poétique" ["Ars Poetica"] "De la musique avant toute chose" ["Music before everything"], advocating a poetry of subtlety and nuance that is as elusive and intangible as the scent of mint and thyme on the morning wind. Similarly, Darío in "Dilucidaciones" ["Elucidations"], his introduction to *El canto errante* [*The Wandering Song*], asserted the affinity of poetry and music and expressed the need for a constantly revitalized and perfectly adaptable poetic form. He wrote: "I do not like either new or old *molds*. . . . My poetry has always been born with its body and its soul, and I have not applied any type of orthopedics to it. I have indeed sung old airs; and I have always wanted to go toward the future under the divine command of music—music of ideas, music of the word" (*Poesía* 304).[25] Martí had made a similar disclaimer with regard to the unconstrained and uncontrived nature of his verse in his introduction to *Versos libres* [*Free Verses*]. He wrote: "Not one has come from my mind reheated, contrived, recomposed; but rather like tears come from the eyes and blood comes in gushes from the wound" (95).[26] This longing for an unfettered, fluid, musical language reflects a view of literature that, as suggested earlier, reaches beyond the aesthetic into the realm of the epistemological and the political. This view of literature becomes the foundation of the *modernista* project.

It is not surprising that Martí, who had resided in New York City for many years and who therefore had confronted "bourgeois modernity" in all its fury, was one of the first writers to envision the imaginative power of this fluid response to modern life. For Martí, living abroad and working for an independent Cuba, the issues of modern life, national identity, and the role of the artist-intellectual became particularly intense. Seeking not only an ideal political, philosophic, and ethical stance but also a means of surviving within an alien and alienating context, Martí tied socioeconomic and literary factors together and proposed not only a truer way of knowing but also an antidote to the excesses of modernization and North American hegemony. If knowledge brings strength, superior perceptivity and consciousness could balance the odds for the less powerful Spanish American nations. In Martí's writings, this merger of poetics and politics is infused with a transcendental impetus appropriate to shaping national identity.

Perhaps the earliest example of this basic *modernista* tendency to conceive of literary and political concerns as two sides of the same proverbial coin appears in Martí's famous prologue to Juan Antonio Pérez Bonalde's *El poema del Niágara* [*Poem of Niagara*]. It was written in New York in 1882 and deserves careful attention. In it Martí confronts the impact of modern life on literary production and defends poetry despite its intrinsic insignificance within the capitalist scheme. He establishes an opposition between "ruines tiempos" ["vile times"] and the spirit of the poet. The images of modern times are easily recognizable: an emphasis on material accumulation and fashionability, a tendency toward vulgarization, and a loss of ideals and idealism. Martí's response grows out of the age-old dialectic between nature and society, but it reflects the subtle tension between modernity and those aspects of life that he hopes to salvage from the all-encompassing impact of "progress." His language is most revealing:

> Under the pretext of completing the human being, they interrupt him. No sooner is he born than they are already standing beside his cradle with great and strong bindings prepared in their hands, the philosophies, the religions, the passions of their fathers, the political systems. And they tie him and they girdle him, and man is already, for his whole life on earth, a bridled horse. . . . One comes into life like wax, and fate empties us into premade molds. Created conventions deform true existence, and true life becomes like a silent current that slips, invisible, beneath the feigned life, not always felt by the very one in whom it works its holy deed, in the same way that the mysterious Guadiana River follows a long path silently beneath Andalucian lands. (111)[27]

We see here a sophisticated view of the emergence of selfhood within the socialization process, which is based on the dichotomies between culture and nature, between rigidity and fluidity, between bondage and freedom.

From the earliest age, from the very first days of existence, one is molded, shaped, and—worse still—deformed by outside forces. These forces are represented as coercive (great and strong bindings) and conservative ("las pasiones de los padres"), imposing a vision from the past. The aggressive and destructive nature of this process is further emphasized by the verbs chosen: "they tie him," "they girdle him." Possibilities and options are channeled along predetermined paths; the individual's being, thinking, and behavior are restricted. The self is poured into previously established molds, made to conform, left with no choices. One might assume that, as a result, the self loses all sense of individuality and simply functions as an adherent of existing social arrangements and belief systems, unable to see the world or act upon it for itself. But regardless of how modern Martí is, he is most definitely not postmodern, and he adamantly defends the concepts of self, identity, and agency.

The true self eludes the absolute oppression of these outside forces by slipping invisibly beneath the surface, which Martí calls "la vida aparente" ["the apparent life" or "the feigned life"]. What is true and real is, therefore, different from what appears on the surface. Martí goes on to compare the underlying current with the mysterious Guadiana River, which flows silently beneath Andalucian lands. Remarkably, for Martí the true self is not destroyed but finds freedom in a hidden, subterranean region. Martí thus affirms the everlasting ability of the self to assert its true nature, one that is portrayed as fluid and unconstrained but also as hidden, as unseen, as, perhaps, inaccessible within the daily routines.

Though Martí's focus in this section is on the self, his entire presentation underscores the epistemological implications that are central to his and most *modernista* writings. Not only is the self prevented from being true to itself, but it is also prevented from perceiving and responding to the world around it in an unobstructed manner. The double meaning of the word *venda,* both bindings and blindfold, highlights this overlap. Because the self is bound by religions, philosophies, passions, and politics, it is unable to achieve an undistorted view of the outside world. In contradiction to the dominant bourgeois culture, which operates on the assumption that outside reality can be measured and assessed and that rational judgments can be made on those measurements, Martí presents a view of self that implies a profound distrust of such naive empiricism; he presents a view that shows his sympathies with the radical philosophical shift that occurred in the nineteenth century and that forms the foundation for philosophical positions that run throughout the twentieth century.

As Henry Aiken has pointed out, before the nineteenth century philosophers "did not, on the whole, seriously doubt that there is a common, independent, and objective reality which can to some extent be understood. Nor did they question whether there is an objective way of thinking about reality, common to all rational animals, which does not radically modify or distort the thing known. Actually they did not deeply ponder the concept of objectivity itself; they merely used it to express a half-conscious conviction about the adequacy of the rational faculty to grasp its object and the correspondence between the thing itself and the thing-as-known" (14–15). He goes on to note that "[f]rom Kant on, however, the assumption of a preordained correspondence between the mind and its object was regarded as dogmatic and uncritical" (15). Thinkers became aware that every conception of reality presupposes a way of thinking about the world and affects what the world is understood to be. Historical consciousness went a step further by acknowledging that not only human nature but also reason develops within history and is continually affected by the changing conditions of individual and social life (15–16).

Whether or not he was introduced to these ideas by German philosophers such as Hegel or Marx, Martí demonstrates a receptivity to this "new" way of thinking. In an 1883 article, "On the Death of Karl Marx," Martí wrote: "Karl Marx studied ways to place the world on new foundations. He awoke the sleepers and showed them how to cast down the broken pillars" (qtd. in Aguilar 103).[28] This breaking down or breaking out is both epistemological and political for Martí, as it was for Marx. Martí's defense of fluidity, spontaneity, mystery, and individuality reveals his distrust of molds, structures, and conventions and, by implication, the visible, measurable, material, and mass-producible. In a later section of Martí's prologue to *El poema del Niágara,* he himself underscores the fundamental connection between poetry and politics.

> To assure human free will; to leave to the spirits their own seductive form; to not tarnish virgin dispositions with the imposition of another's prejudices; to ready them to take what is useful for itself alone, without confusing them nor pushing them along a marked route. . . . Neither is there room for literary originality nor does political freedom subsist as long as spiritual freedom is not assured. Man's first task is to reconquer himself. (111)[29]

While Martí's central concern with struggle and renewal is clearly evident in this final militaristic image, the source of his militancy is hidden within the alternative way of knowing predicated on the ability to imagine the world anew.[30] He proposes a reconquest of self and society armed with a different type of power. He assumes a moral and political superiority that is derived from a vision that resists the limitations of preset cognitive structures and that relies upon a poetic and premodern mode of understanding.

Even though he does not elucidate the exact nature of this alternative way of knowing, Julio Ramos agrees that it is a source of power and authority. For Ramos it becomes the means by which "literature begins to authorize itself as an alternate and privileged mode to speak about politics. Opposed to the 'technical' ways of knowing and to the imported languages of official politics, literature presents itself as the only hermeneutics capable of resolving the enigmas of the Latin American identity" (16).[31] Using Marxist terminology, he points out literature's ongoing struggle with "bourgeois modernity." He writes: "Its [literature's] economy will be, at times, a way of granting value to materials—words, positions, experiences—*devalued* by the utilitarian economies of rationalization" (10).[32]

Of course this "privileging" of the literary voice in the cacophony that filled the political debates of the day is made possible by asserting literature's grounding in a much older, "truer" way of knowing.[33] It is the imagery based on analogy that becomes the foundation of the *modernista* epistemology as well as its challenge to bourgeois values. The premise that nature holds a hidden system of correspondences that reveals a divine and harmonious order toward which man must be free to aspire becomes the *modernista* answer to the stultifying rules of Spanish poetics. More importantly, however, it supplies a satisfying response to the modern world—to facile assumptions about science, scientific knowledge, and the unexamined positivist pursuit of progress. By extension, analogy offers an answer to North American hegemony as well, one in which the values of democracy can be praised without resigning Spanish American reality to a second-class status.

Though Martí is the *modernista* most widely recognized for his political activism, he was by no means the only one to perceive a political dimension to the struggle for poetic freedom. In one of the most direct statements on the subject, another early *modernista*, the Mexican Manuel Gutiérrez Nájera, explains how the formation of an expansive, open, and receptive literary discourse is understood by established, conservative elements of society to reflect political aspirations that threaten the literary and political old guard. Attempts to open the way for a new literary language, to break the strictures of inflexible rules, and to expand the realm of the sayable are tied to the struggle against the conservative, intolerant, and self-perpetuating establishment. In "La academia mexicana" ["The Mexican Academy"], the connection is made explicit: "Until now, with only a few exceptions, the Academy has been made up of persons addicted to the throne and to the altar; [made up] of men fearful of God and grammar, who with equal rectitude oppose sins against the law of God and sins against orthodox syntax. . . . The slightest liberal inconstancy, the most trifling carelessness in syntax, a *le,* a *lo,* a sonnet to Juárez, is enough to close the sanctuary of vowels and consonants to the candidate" (*Obras* 248). Gutiérrez Nájera's complaint is simple: "What I am censuring is their intolerance—more political than philosophic" (*Obras* 259).[34]

In another article, "El arte y el materialismo" ["Art and materialism"] (*Obras* 49–64), Gutiérrez Nájera juxtaposes realism and sentimental poetry, which he defends not only for its aesthetic worth but also for its expression of patriotic and social concerns. He writes: "Guided by a highly spiritual and noble principle, encouraged by a patriotic, social, and literary desire, focusing on elevated goals, we raise our humble and weak voices in defense of sentimental poetry, so many times trampled, so many times attacked, but triumphant over low-spirited theories of realism and over disgusting and repugnant positivism" (*Obras* 50).[35] In order not to leave any doubts about his meaning, he goes on to clarify the political underpinnings of his literary agenda. He relates the struggle for beauty, goodness, and truth with the fight for freedom and for the privilege of envisioning realities beyond those of established political and poetic structures and molds. He makes this connection by happily announcing that realism has found few advocates among Spanish American lyric poets and by proclaiming: "And how should it not be so, if we, children of ardent America, brave soldiers of liberty, born in the beautiful valleys where Spring has its eternal rule, hate all servitude, break all chains, and loving with infinite love all that is true, good, and beautiful, let our imagination fly freely and give free run to all our noble sentiments!" (*Obras* 63).[36] Martí's often repeated militaristic images find an antecedent in this 1876 piece as Gutiérrez Nájera conjoins hard-won literary and political freedoms.

At the other end of the hemisphere and nearly thirty years later, the Argentine poet Leopoldo Lugones also speaks of the power of poetic language as a national asset, one that must be cultivated and cared for. In his prologue to *Lunario sentimental* [*Sentimental Lunar Calendar*] (1909), he writes: ". . . thus to find new and beautiful images, expressing them with clarity and concision, is to enrich the language, renovating it at the same time. Those in charge of this task—at least as honorable as that of breeding cattle or of administering state revenues, since it is a question of a social function—are the poets. Language is a social resource, perhaps the most consistent element of national manners and customs" (192).[37] Despite the verbal artifice and fireworks of collections like *Las montañas del oro* [*The Mountains of Gold*], *Los crepúsculos del jardín* [*Garden Twilights*], and *Lunario sentimental,* Lugones never loses sight of the relationship between the formation of nation-states and the creation of a modern mode of discourse appropriate to them. Civil concerns come into sharp focus in *Odas seculares* [*Secular Odes*], a collection published in 1910 as part of the centennial celebration of Argentina's independence.

Lugones's early works are not the only ones in which the political dimension to *modernismo* lurks behind what appears to be purely literary concerns. In "El Modernismo," the Mexican poet Amado Nervo reveals his impetus to create a poetic language that is capable of reflecting, on one hand, "modern" attitudes and desires and, on the other, the changing Spanish American scene. "We did not have words to say the new things that we see and feel; we have looked for them in all the dictionaries, we have taken them, when they were there, and when not, we have created them" (2:299).[38] As a result, he offers no apologies for his rejection of "the old grammatical combinations, the old phonetic arrangements," which had become set phrases, ritualistic patterns of speech (2:399).

While Nervo justifies this break with tradition, in another article, "El castellano en México" ["The Castilian of Mexico"], he defends the Spanish spoken in his homeland and rejects the criticism of certain *modernista* vocabulary as Gallicisms. With an extensive and carefully formulated list, Nervo shows that the Spanish spoken in Mexico is not incorrect; it merely seeks to resurrect older forms of speech. He shows that, when Spanish American *modernistas* used the more Latinate words for "goldsmith" or "young woman" or "malicious," they did not resort to Gallicisms "but rather simply disinterred words that had fallen into disuse without any reason, since they were beautiful, as the case of the first two, or they did not have exact substitutes, as the last one" (2:103).[39]

Nervo's assertive cosmopolitanism and innovativeness in one article is balanced by a more traditionalist defense of Mexican and *modernista* Spanish in another. This juxtaposition underscores the complexity of the *modernista* attitude toward language and its political repercussions. Not content to follow well-beaten paths or to repeat hollow clichés, *modernistas* struggled to create a language that could both shape and reflect a Spanish American identity still in formation.

This conception of literature as a means of attaining and disseminating knowledge—metaphysical and political—cannot be emphasized enough, for it dispels, once and for all, any lingering doubts about the seriousness of the *modernista* project and clarifies the intricacy of the issues with which the *modernistas* sought to grapple. Their understanding of the importance of their task developed, as might well be expected, within a context in which language was not a simple issue. The most public and widely studied debate on this topic began in 1842 and centered on questions of language, national identity, and the political direction of the newly independent countries of Spanish America. The controversy aligned Domingo Faustino Sarmiento and José Lastarria on one side and Andrés Bello on the other. The politics are complex, for they include issues of national pride (the arrival in Chile of Sarmiento, Juan Bautista Alberdi, and Vicente Fidel López, all of them Argentines), political orientations (the intense conflict between conservatives and liberals), and generational control (Bello, the senior intellectual, was thirty years older than Sarmiento).[40]

Sarmiento began the polemic with his caustic review of a collection of essays by Pedro Fernández Garfias entitled *Ejercicios populares de la lengua española.* The goal of Garfias's work was to correct the misuse of the Spanish language by Chileans. Sarmiento's criticism of his reactionary position was later elaborated in a seventy-page treatise that proposed a less traditional and more spontaneous use of language. Sarmiento held that the solid but rigid humanistic formation of the younger generation of Chilean writers actually worked against their productivity and made them incapable of creating poetry. Emir Rodríguez Monegal astutely summarizes the argument:

> His [Sarmiento's] thesis, romantic in origin, is that a language is the expression of the ideas of a people, and a people has to take its ideas where they are, independent of the criterion of linguistic purity or academic perfection; that Spanish literature has lost all its strength and that America is no longer willing to wait for foreign ideological trading to pass through Spanish heads in order to be able to partake of it; that the real function of the Spanish Royal Academy is to collect, as in a clothes closet, the words that people and poets use and not to *authorize* the use of the same; that languages return today to the people; . . . and that the influence of grammarians, the fear of rules, and the respect for admirable models have the Chileans' imagination throttled. (264–265)[41]

Bello appears in these exchanges as the representative of neoclassical inflexibility, even though he proved to be more open and understanding than he has been made out to be. Bello's vision was shaped by his longing to establish cultural reunification with Spain based on the conservation and enhancement of Spanish America's Iberian heritage. He believed that the cultivation of language and literature would lead to solidarity throughout the Hispanic world. For his defense of Spanish culture he was attacked by one of his own students, José Lastarria, who held that Spanish America had to free itself of the undemocratic and retrograde cultural domination of Spain. As Efraín Kristal asserts, Sarmiento went even further. "Sarmiento argues that Hispanic America must break with the Spanish tradition because Spain is unable to express modern ideas. He says that the Spanish inability to break with established norms impeded the artistic development of Hispanic Americans. He points to France as a superior source of ideas and blames Bello's teachings for the intellectual and artistic poverty of Chile" ("Dialogues" 65).

Bello responded in 1843 in his speech at the opening of the University of Chile, where he served as the first president. While he remained firm in his advocacy of cultural continuity, he was not intractable. He accepted the possibility of growth and change within the established linguistic frameworks: "But language can be expanded, it can be enriched, it can be accommodated to all the demands of society and even of fashion, which exercises an undeniable sway over literature, without adulterating it, without corrupting its constructions, and without doing violence to its nature."[42] Said another way, his goal was to have the youth of Chile think for themselves, write well, and be creative but not lose themselves in excesses that would transform the language beyond recognition.[43]

More specifically, he was not prepared to accept all that Sarmiento and Lastarria proposed, that language and literature should be left in the hands of the general public. Julio Ramos, relying upon Sarmiento's own words in his defense of *Facundo,* focuses upon this point of disagreement and underscores how these issues are tied to the question of knowledge and politics.

> The task of the "poor American narrator" could perhaps wind up being "undisciplined" or "irregular" (attributes of barbarousness). But that "spontaneity," that nearness to life, that "immediate" discourse was necessary in order to represent the "new world" of which European knowledge, despite its own interests, was ignorant. [. . .] for Sarmiento one had to know that whole area of American life—barbarousness—that wound up being *unrepresentable* for "science" and "official

documents." It was necessary to hear the *other;* to hear his voice, since the other did not have writing. That is what disciplined knowledge, and its importers, had not managed to achieve; the *other* knowledge—knowledge of the other—would wind up being decisive in the restoration of order and of the modernizing project. (24)[44]

He stresses that "Sarmiento insisted precisely on the extrauniversity formation of his discourse, spontaneous and even undisciplined, but for this reason more able to understand the American 'barbarousness'" (35).[45]

This position reveals the origins of issues that take center stage in *modernismo.*[46] The search for a language that can reflect the needs and wants of the writers and thinkers struggling against the narrow, rigid precepts of the ruling establishment becomes the foundation of *modernista* goals. The mission is to create a spontaneous, natural, fluid, intuitive language that is truer, more authentic, more representative of the native spirit than that which has been imported from Spain, imposed by neoclassical rules, or dictated by the unthinking old-guard. This new language would not focus on the external realities of the positive sciences but would turn to the subtle realities of the soul—the national essence and the individual being. *Modernismo*'s success, its ability to actually alter the direction of Spanish literature, owes a debt to the earlier debate.

At the same time, however, the grounding of the old debate changed with the introduction of modernity to Spanish American life. Clearly the end of the colonial period and the first years of independence brought changes that are linked to the transition to modernity. Carlos Fuentes, for example, now speaks about Spanish American independence in these very terms.

Spanish America broke away from Spain between 1810 and 1821, and the break was not only political, but moral and aesthetic as well. For most liberals, independence meant renouncing the Spanish past as a reactionary, intolerant, and unscientific era of darkness. It meant promptly attempting to recover lost time by achieving all that the Spanish Counter-reformation denied us: capitalism, free inquiry, free speech, due process, parliamentarism, industry, commerce: in short, modernity. (Prologue, *Ariel* 15)

While independence at the beginning of the nineteenth century opened the floodgates of modernity, the *modernistas* were the first writers to live with and appreciate the all-encompassing transformations that were altering the nature of life in Spanish America. Industrialization, mechanization, commodification, international trade, and the positivistic emphasis on science, reason, and material prosperity only began to permeate life in the capitals of Spanish America during the second half of the nineteenth century. The crisis that these changes generated shaped the essential character of *modernismo* and defined its relevance for later writers. The various ways this character developed among the early *modernistas* is the focus of the next chapter.

THREE
The Movement Takes Shape

THE *modernista* movement started simultaneously in several parts of the continent during the last quarter of the nineteenth century. It began as an outgrowth of romanticism, which nurtured the development of an original literature for the recently independent countries of Spanish America. *Modernismo* reacted to and was, at the same time, sustained by the affluence and cosmopolitanism of urban centers such as Mexico City, Havana, and Bogotá in the north and Santiago de Chile, Buenos Aires, and Montevideo in the south. These emerging metropolises provided support for the most important literary magazines of the Spanish-speaking world, magazines that were crucial in the diffusion of concepts, trends, and translations which led to the formation of a continent-wide movement. The most famous of these journals were *Revista Azul, Revista Moderna, El Cojo Ilustrado, El Mercurio de América,* and *Vida Moderna.*

The qualities that characterize what would come to be known as *modernista* writings appear in the earliest works of the movement. To varying degrees one notes a general dissatisfaction or disillusionment with the values of the ruling class, anxiety and a sense of crisis with regard to traditional religious beliefs, and an eagerness to establish a new, uniquely Spanish American perspective. The first *modernistas* aspired to produce a poetry that would rival the esteemed art and imported items that were entering the great cosmopolitan centers of Spanish America from all corners of the world, but most especially from Europe and Europe's cultural capital, Paris. Their poetry sought to liberate verse form from the inflexibility of classical norms and to achieve a powerful vision based on an innovative conception of language, one in which musicality and formal perfection were thought capable of evoking profound realities.

By the 1880s the first *modernistas* had produced a fundamental transformation in Spanish literature. It was Darío, however, who became the head of the movement, who gave it its name, and who propelled it forward with driving energy and genius. By the end of 1896, the year in which he published the first edition of *Prosas profanas,* a work of revolutionary vision and artistry, he stood alone as the undisputed head of and spokesman for *modernismo.* Because of an unfortunate turn of events, by the end of that year, the best early *modernistas,* Manuel Gutiérrez Nájera, José Martí, Julián del Casal, and José Asunción Silva, were dead, and the others had all written their best poetry.[1]

Born and raised in Mexico City, Manuel Gutiérrez Nájera was, as noted in chapter 2, one of the first *modernistas* to recognize that the stultification of Spanish verse stemmed not from inherent linguistic limitations but rather from a resistance to change fostered by conservative factions, some of which were aligned with the political old guard. He advocated greater artistic freedom that was consistent with the cultural openness evident in his beloved Mexico City and in his famous poem "La duquesa Job" ["Duchess Job"]. The title of the poem alludes indirectly to his pseudonym "El duque Job," the most famous of the approximately thirty that he used during his career as a writer. In this revealing piece, the imported models of elegance and grace which Mexican society was beginning to emulate were never proposed as substitutes for uniquely Spanish American attributes. Though he dubs his beloved "la duquesa Job," thereby playing with references to European nobility and literature, and though he fills the poem with foreign vocabulary and allusions, Gutiérrez Nájera affirms the superiority of the one that can navigate among the most attractive features of other cultures while remaining true to oneself.[2] He ends the poem with the following stanza:

> Desde las puertas de la Sorpresa
> Hasta la esquina del Jockey Club,

No hay española, yankee o francesa,
Ni más bonita ni más traviesa
Que la duquesa del duque Job! (2:24)

[From the doors of the Surprise / to the corner of the Jockey Club, / there is no Spanish, American, or French woman, / prettier or more mischievous / than the duchess of Duke Job.]

His enthusiasm for European literature and his advocacy of beauty, musicality, and art for art's sake reflected his faith in artistic expression to reveal profound realities. He hoped that these realities would fill the spiritual void left by the vitiated discourse of traditional religion and by the materialistic ideology of the "modern" businessman and scientist. As he repeatedly made clear in his extensive critical writings, he believed that the intuited and evocative wisdom of art would benefit the public good as well as serve national ends. As a result, the Gutiérrez Nájera that has been identified as "afrancesado" ["Frenchified"] cannot be judged superficially. His playful cosmopolitan spirit, his poetic experimentation (as in the Parnassian "De blanco" ["In White"] or "El hada verde" ["The Green Fairy"]), and his search for the perfect adaptation of image and language and of color and tone are all products of serious reflection upon both the promises and failings of the expanding cultural and commercial milieu of the time.

Gutiérrez Nájera knew the changing Mexican scene well. As a professional journalist, he kept his readers, many of them women, informed of the events, trends, and fashions of the day. He wrote reviews, commentary, chronicles, and short stories. He actually produced much more prose than poetry, which was collected only posthumously by Justo Sierra in 1896 in a two-volume edition entitled *Poesías* [*Poems*]. Nevertheless, his poetry was well known and quite influential during his lifetime. It was admired for the vitality and elegance it brought to Spanish verse, even though his technical innovations were limited to the introduction of new accentual patterns within traditional metrical forms, especially the octosyllable and the hendecasyllable. While Gutiérrez Nájera's confrontation with the spiritual abyss left by the materialistic and positivistic perspectives dominant at the time lends a melancholy and tortured air to much of his poetry, it is its lilting musicality and delicate imagery that leave the greatest impact.

A perfect example of these features is "Para entonces" ["By Then"], the first poem in his posthumous *Poesías.* This composition of four elegantly constructed quatrains *(serventesios)* captures a profound longing to escape from the world of human interaction and suffering.

> Quiero morir cuando decline el día,
> en alta mar y con la cara al cielo;
> donde parezca sueño la agonía,
> y el alma, un ave que remonta el vuelo.
> No escuchar en los últimos instantes,
> ya con el cielo y con el mar a solas,
> más voces ni plegarias sollozantes
> que el majestuoso tumbo de las olas.
> Morir cuando la luz, triste retira
> sus áureas redes de la onda verde,
> y ser como ese sol que lento expira:
> algo muy luminoso que se pierde.
> Morir, y joven: antes que destruya
> el tiempo aleve la gentil corona;
> cuando la vida dice aún: soy tuya,
> aunque sepamos bien que nos traiciona! (1:29)

[I want to die when day approaches its end, / at high sea and with my face toward the sky; / where agony appears a dream, / and the soul, a bird that takes to flight.

Not to hear in the last moments, / already alone with the sky and the sea, / other voices or tearful prayers / than the majestic rhythm of the waves.

To die when the light, sad, withdraws / its golden nets from the green wave, / to be like that sun that slowly expires: / something very luminous that is lost.

To die, and young: before treacherous / time destroys the gentile crown; / when life still says: I am yours, / even though we well know that it betrays us.]

The poem's structure and images set up a contrast between the grace and dignity of the natural order and the misery resulting from the human condition in modern society. Unmentioned are the urban chaos and congestion from which the voices and sobbing prayers seem to emanate and from which the "I" of the poem seeks to withdraw. Yet the desire to die alone, on the open sea, where suffering no longer seems real, asserts the hateful nature of the shore—and life on it. Peace is measured by distance from land and the social and commercial exchanges on it.

The sea and the rhythms of nature offer what human company cannot: serenity, dignity, and hope. For this reason, the speaker expresses the desire to link his fate with that natural order, finding solace in the patterns found there. By the third stanza, death appears as part of the harmonious accord of the universe. It is as correct and appropriate as life itself is not. Consequently, the wish to die young of the fourth stanza is more than a rejection of old age. It is a renunciation of the cruelty that precedes or accompanies it. In particular, the personification of life refocuses the poem on human interactions, such as those deriving from erotic love or fraternal commitment, as well as on the pain that inevitably results from them.

This poem expresses a longing to live—and die—in contact with the eternal truths rather than the ephemeral and treacherous circumstances of personal relationships. The ultimate embrace of death, not as the long-awaited completion of the life cycle but rather in the full bloom of youth, is exemplary of the romantic and *modernista* rejection of society that, because of its dominant values and structures, betrays those that care about profound truth and eternal beauty.

Gutiérrez Nájera's confrontations with the disappointments of the human enterprise, the destructive passage of time, and the overwhelming presence of death are repeatedly converted into masterpieces that underscore not only the redemptive power of art, as in "Non omnis moriar," but also his unending search for beauty, beauty that could permeate human actions and influence moral behavior. This desire—rooted in Hegelian philosophy and German romantic idealism—to influence, through his art, the course of his nation's history shows that the *modernista* movement began with full recognition of the breadth and seriousness of its quest. This early intensity of purpose stands out even more in the works of José Martí, often considered the most imposing figure of this period.

It has been virtually impossible to speak of Martí without alluding to his dedication to the cause of Cuban independence, which formed the intellectual backdrop to his political, moral, and philosophic writings and which ultimately led to his death on the battlefield at Dos Ríos. His prose pieces (particularly his speeches), with their powerful and innovative use of the Spanish language and their insightful reflections upon the developments at the turn of the century, have tended to receive more attention than his three volumes of verse. His poetry, as he himself confesses, served as a break from his primary tasks of journalism and politics. Nevertheless, those features that have been singled out for praise in his prose are also essential to his poetry, which constitutes a fundamental component in the formation of the *modernista* movement. Among the most important aspects of his writing are his strong sense of moral imperative and his insistence that language conform to the lyrical impulses that drive it—even at the risk of being shocking or brutally sincere. As Cintio Vitier has pointed out in his "En la mina martiana" ["In the Martian Mine"], Martí not only sought to make a revolution through his words but he also hoped to revolutionize language itself, making it more American, receptive to modern times and, perhaps more importantly, relevant to the future. His famous phrase "La expresión es hembra del acto" ["The expression is the female of the act"] underscores the bond between word and deed, suggesting, like the female metaphors that run throughout Darío's work, that language, upon being fertilized, should give birth to verbal children, who, in Martí's case, are the moral equivalent to the deeds of just men.

It was possibly this ethical orientation, combined with Martí's lasting respect for traditional Spanish verse forms—despite his enthusiasm for the French and North American literature of the day, most notably the visionary figure of Hugo, the sonorous cadences of Whitman, and the oneiric fantasies of Poe—that made the movement's first critics slight his contributions to *modernista* verse. Moreover, with the early but persistent misconception that *modernismo* was basically an aesthetic movement unconcerned with the "real" world, Martí appeared anomalous, hard to explain.[3] However, as a number of more recent studies have sought to place *modernismo* in the context of modernity in general and the formation of modern nation-states in particular, Martí's place within the literary movement has become progressively clearer.[4]

In effect, Martí's literary innovations address the essence of modern Spanish America and thereby anticipate not only *modernista* struggles with language and society but also those of the avant-garde. Martí's willingness—perhaps eagerness—to tap the power of dreamed, unfettered, and even illogical visions and verbal structures is picked up later by Julio Herrera y Reissig and Leopoldo Lugones and underpins what Yurkievich has called the "causal connection between *modernismo* and the first *vanguardia.*"[5]

As noted in the second chapter, Martí's writings offer some of the earliest examples of the *modernista* tendency to envision literary and political concerns as one and the same. Martí's own analysis of his poetry in the prose introduction to his *Versos libres* [*Free Verses*] (completed in 1882 and published posthumously in 1913) underscores the union of these perspectives within an analogical worldview. He wrote:

> These are my verses. They are as they are. I borrowed them from no one. Since I could not entirely enclose my visions within a form adequate to them, I let my visions fly. Oh, how many a golden friend that has never returned! But poetry has its integrity, and I have always wanted to be honest. I also know how to clip verses, but I do not want to. Just as each man brings his own physical attributes, each inspiration brings its own language. (95)[6]

Martí's unswerving commitment to have his vision choose its poetic form, his metaphoric brilliance and novelty, and his acceptance of the difficulty of his verse—along with its pure, often brutal, sincerity—are characteristics that establish his work within the ancient tradition of analogy that was resurrected by the romantics and symbolists. This faith in the power of sincerity that surfaces in Martí's writings (as well as in Darío's and that of other *modernistas*) can only be understood in this context.

Sincerity is the purest, least contrived expression of the natural concordance between the poet and the world, and it is the goal that Martí sets for his poetry. Integrity [*honradez*] must be contrasted, as Roberto González Echevarría has pointed out, with decorum [*decoro*]. *Honradez* emphasizes the need for poetry to be faithful to the spontaneous and unfettered nature of the poet's vision. *Decoro,* on the other hand, is the classical concept that alludes exclusively to a faithfulness to established poetic norms ("Martí" 31). With his selection of the word *honradez*, rather than the often used *sinceridad,* Martí stretches, once again, what is a predominantly poetic concept for others into a moral imperative, one that ultimately demands more than verbal honesty. It points to a life that must be lived with integrity, in accord with a more perceptive vision of the natural and social order.

While *Versos libres* was completed during the same year as *Ismaelillo* [*Dear Little Ishmael*] and contains poems from as early as 1878, *Ismaelillo* is Martí's first published book of verse. It consists of fifteen poems dedicated to his absent son, José Francisco (Pepito), who had been born in 1878. The personal events surrounding their separation and the creation of these poems underscore how thoroughly Martí's commitment to the cause of Cuban independence permeated his life and work. While he was living in New York with his wife, Carmen Zayas Bazán, she often accused him of caring more for Cuba than for his family. One day, without notice, she left for

Cuba with their son, whom Martí would never see again. *Ismaelillo* grew out of the pain of this loss. In contrast with the texts of the short-lived journal *La edad de oro* [*The Golden Age*] (1889), this book is not for children but rather about one. It captures the pure, spontaneous joys of parenthood as well as Martí's hopes for his son and for the future, which come together in the sense of mission and purpose that he aspires to pass on. It is this tension between the lyrical innocence of the child and the moral world of the father that structures the work and also makes it appear so fresh, dynamic, and modern. The son becomes a knight, a shield, a refuge for the father. Through these images the work fuses fantasy, spiritual comfort, and moral action. For this reason, Santí believes that the title of the collection alludes to the etymology that Martí attributed to Ishmael, namely, "ser fuerte contra el destino" ["to be strong against destiny"]. The child is imagined to be heroic despite his not having had to suffer the trials of exile like his true father or his Biblical namesake who, along with his mother, was cast out by Sarah, Abraham's legitimate wife.

Despite the fact that *Ismaelillo* was too little read during Martí's life to have been influential, its style reflects and represents the best features of early *modernismo.* It makes maximum use of the traditional *seguidilla,* giving it a light and energetic air.[7] Verbs are chosen for their sense of movement, nouns for their metaphoric power, and adjectives for their pictorial qualities. The overall impression is one of activity that borders on chaos, a chaos that reinforces the urgency of the emotions that well up uncontrollably within the loving father.

This sense of urgency also dominates the contemporaneous *Versos libres* and the poems of Martí's next collection, *Flores del destierro* [*Flowers of Exile*] (written between 1882 and 1891 but published posthumously in 1933). In both works, there is a verbal abundance, a volcanic flow of emotions and images that suggest the directness and spontaneity of drafts rather than the studied revisions of polished pieces. The adjectives are strong and often surprising, and the verbs usually appear in the present, the infinitive, or the imperative, communicating action and activity. Though these works address the same anguish and struggles as *Ismaelillo,* their focus is more universal and the perspective more existential, once again projecting a kinship with the poetry of the twentieth century and asserting an unexpected affinity with the avant-garde.

All the tensions and pressures that he felt as a committed, conscientious individual are verbalized in what he called the "endecasílabos hirsutos" ["rough, hairy hendecasyllables"] of these poems. With a clearly stated confidence in the harmonious order of nature, these works explore the themes of exile, love, loss, justice, responsibility, and poetry. Anticipating images similar to those of Darío's "Era un aire suave . . ." ["It was a gentle air . . ."] and "Yo soy aquel que ayer no más decía . . ." ["I am the one who only yesterday was saying . . ."], Martí in "Poética" ["Poetics"] describes his verse as capable of going to the salons of the rich and royal, capable of courting ladies and princesses. Yet his poetry prefers the silence of true love and the dense growth of the jungle. There, and not in the urban centers of imported and superficial values, the poet can read the secrets of the universe that he must decipher for his readers. His rejection of the city is complete, for to him it represents hypocrisy, fakery, and the masses of poor and suffering. In contrast, the countryside represents purity, sincerity, and contact with the universal forces that give meaning to all the mysteries of life—including love and death. Yet Martí knows, as he makes clear in "Amor de ciudad grande" ["Big City Love"], that the present and future belong to the city, to the world contaminated by time and sin.[8]

His distaste for the urban world could not be any more apparent. The second half of the first section underscores the key issues of modern life:

> ¡Así el amor, sin pompa, ni misterio
> Muere, apenas nacido, de saciado!
> ¡Jaula es la villa de palomas muertas
> Y ávidos cazadores! ¡Si los pechos
> Se rompen de los hombres, y las carnes
> Rotas por tierra ruedan, no han de verse
> Dentro más que frutillas estrujadas! (125)

[In this way love, without pomp or mystery / dies, just born, of being satiated! / The town is a cage of dead doves / and avid hunters! If men's chests / break, and the broken flesh / rolls over land, one will not see / anything more than squashed little fruits!]

Love dies unable to maintain its mystery through innocent and unfulfilled longings. The city is geometrized; its paths and structures metaphorically converted into bars and barriers. People are dehumanized and destroyed. The dispensability of individuals and the loss of their human identity and dignity generate an image so powerful and modern that it reappears nearly seventy years later in the final section of "La United Fruit Co." from the fifth section of Pablo Neruda's *Canto general.*

Mientras tanto, por los abismos
azucarados de los puertos,
caían indios sepultados
en el vapor de la mañana:
un cuerpo rueda, una cosa
sin nombre, un número caído,
un racimo de fruta muerta
derramada en el pudridero. (336)

[Meanwhile, in the seaports'
sugary abysses,
Indians collapsed, buried
in the morning mist:
a body rolls down, a nameless
thing, a fallen number,
a bunch of lifeless fruit
dumped in the rubbish heap.

(Translation by Jack Schmitt 179)]

Like Neruda (and Vallejo as well), Martí struggles with the obvious limitations of artistic involvement in the face of devastating social changes. He finds consolation in a purity of purpose that reflects the orderly perfection of nature. All these elements come together in Poem XVII of *Versos sencillos* [*Simple Poetry*] (1891), as he envisions his poetry as the musical expression of the loving, knowing rhythm of the cosmos that passes through his soul. The poem ends with the following stanza:

¡Arpa soy, salterio soy
Donde vibra el Universo:
Vengo del sol, y al sol voy:
Soy el amor: soy el verso! (195)

[I am a harp, I am a psaltery / where the Universe vibrates: / I come from the sun, and I go to the sun: / I am love: I am poetry!]

With its recourse to the most common of verse forms in the Hispanic tradition, octosyllables, *redondillas,* and quatrains, this last collection comes closest to achieving Martí's aspiration of clarity, simplicity, and harmony.[9] The work's popular tone and its Pythagorean vision of harmony offer a pristine image of Spanish America and evoke a direct, natural, intuitive wisdom that is easily contrasted with the artifice of Europe and North America. The fusion of art, politics, biography, philosophy, and passion give these simple lyrics a profound and universal transcendence.

Martí's poetic vision is a response to the failings of modern life and a source of hope, but the poet is not immune to the disheartening realities that surrounded him. Indeed, Martí's modernity can be measured, in part, by the irreconcilability of his aspirations with the world that intrudes upon him. "Yo sé de Egipto y Nigricia . . ." ["I know about Egypt and Nigritia . . ."], the second poem from *Versos sencillos,* exemplifies this tension and many other points that I have been making.

> Yo sé de Egipto y Nigricia,
> Y de Persia y Xenophonte,
> Y prefiero la caricia
> Del aire fresco del monte.
> Yo sé de las historias viejas
> Del hombre y de sus rencillas;
> Y prefiero las abejas
> Volando en las campanillas.
> Yo sé del canto del viento
> En las ramas vocingleras:
> Nadie me diga que miento,
> Que lo prefiero de veras.
> Yo sé de un gamo aterrado
> Que vuelve al redil, y expira,—
> Y de un corazón cansado
> Que muere oscuro y sin ira. (181–182)

[I know about Egypt and Nigritia / and about Persia and Xenophon, / and I prefer the caress / of the fresh air of the countryside.

I know about the old stories / about man and his quarrels; / and I prefer the bees / flying among the bellflowers.

I know about the wind's song / among the chattering branches: / let no one say that I lie, / I truly prefer it.

I know about the terrified buck / that returns to the fold and expires,— / and about the tired heart / that dies dark and without anger.[10]]

Martí here sets out to contrast the great ancient cultures with the simple, untouched purity of the countryside. The verb *saber* [to know], which is repeated in the first person singular at the beginning of each stanza, makes the focus of the poem knowledge; the structure of the first two stanzas emphasizes the tension between nature and society and the ways of knowing represented by each. The contrast between "yo sé" ["I know"] and "y prefiero" ["and I prefer"] underscores that the preference expressed is the result of an informed choice. The natural world is more desirable than the cultural achievements of the past, for inherited models are limited and limiting. Histories are filled with quarrels that are presented as eternal repetitions of the same old mistakes: "las historias viejas / del hombre y de sus rencillas" ["the old stories / about man and his quarrels"]. The earthbound failings are contrasted with the flying bees, the stale histories with the fresh air of the countryside, the voices of the past with the song of the wind.

Given the origin of *Versos sencillos,* it would not be unreasonable to see in these oppositions—in addition to the obvious poetic and epistemological implications—less immediately evident metaphoric associations related to nationhood and national identity. In 1888 the government of the United States called the First Conference of American States, which was held in Washington, D.C., between October 1889 and April 1890. In an article entitled "Congreso internacional de Washington" ["International Conference in Washington"], Martí expressed his fear that the conference had been organized by the United States to establish a new type of colonialism from which Spanish America would have to declare its second independence.[11] These worries led Martí to produce *Versos sencillos;* Martí himself states that he wrote these poems as an escape from the anxieties brought on by the meetings.

The patterns from the past that Martí alludes to, therefore, have as much to do with politics as poetic structures or ways of knowing. Those that do not correspond to the authentic nature of Spanish American countries can only be seen as inadequate. It is folly to try to impose a foreign blueprint on the essential character of these nations. While this stance has now gained considerable respectability, it represented an inspired warning to maintain integrity and autonomy in the face of renewed foreign intrusions. Martí's conclusion underscores the terror that one feels upon seeing oneself forced to alter one's nature, forced to conform, forced into the fold, which is literally the word chosen by Martí to describe the manner in which the buck dies. John M. Bennett, in discussing the end of the poem, refers to an ancient hunting practice by which an animal is trapped by continually tightening a net that covers a certain territory. There is no need to belabor the obvious parallels.

At this point in the poem, Martí, unrelenting in his vision, acknowledges his own exhaustion. His is the tired heart. It dies in the dark, unrecognized, resigned to the philosophic and political struggles of modern life. Martí never avoided the "good fight"; he died pursuing its noble goals. Others, however, sought other ways of responding to the new world order.

It is revealing—about the *modernista* movement and the writers in question—that Julián del Casal should in many ways appear to be, as noted by Cintio Vitier, the antithesis of his compatriot and contemporary José Martí (*Lo cubano* 285). These differences stem from different worldviews, life expectations, and temperaments. Martí's faith in man to act morally and to re-create in his life and art the harmonious order of the universe permeated all that he did and wrote. For Casal life was a constant disappointment, flawed by personal indignities, human failings, and cosmic injustices. He resented having to work as a journalist and lost his job repeatedly, but he had few options and was forced to return to writing for newspapers about the social, literary, and theatrical scene in order to earn a living. Throughout his career, he offended many as he struggled against what he considered crass materialism, ignorance, and bad taste. For the greater part of his life he suffered physical pain and, toward the end, he knew that he would die young.

As a result of all these personal as well as other philosophic and artistic factors, Casal saw art not as the epitome of nature's perfection, as it was for Martí, but rather as something different from and superior to nature. Art was the result of human endeavors; its goal was the production of an artificial beauty that more often than not elaborated upon other highly esteemed objects of human creation. The magnificent Parnassian sonnets of "Mi museo ideal" ["My Ideal Museum"] of *Nieve* [*Snow*] (1892), for example, were based on the paintings of Gustave Moreau. In general, Casal's poems draw upon painting, poetry, and crafted materials of all sorts; they evoke exotic and imaginary settings and affirm the human-made environment of the city and its luxurious urban interiors. Casal makes little effort to hide the influence of his predecessors. The anguished sentimentality of the romantics (Zorrilla, Bécquer, Heine, Leopardi), the acute and demanding aestheticism of the Parnassians (Gautier, Heredia, Coppée), the subdued attraction to the unobserved realities of the symbolists (Baudelaire, Verlaine), and the naughty self-indulgence of the decadents (Baudelaire, Huysmans) are openly present throughout the three collections that make up his lifework—in his first two books of verse, *Hojas al viento* [*Leaves in the Wind*] (1890) and *Nieve* (1892), and in his third, *Bustos y rimas* [*Busts and Rhymes*] (1893), which was prepared during his life but published shortly after his death and which contains both prose and poetry.

This exaltation of artificiality is not the only feature that distinguishes Casal's poetry from Martí's. Casal's dissatisfaction with the status quo does not generate, as in Martí, an optimistic, energetic, outward thrust, a push toward change. Quite the contrary, his poetry is marked by an inward turn, the exploration of subtle psychic states, and the presence of suffering, death, boredom, bitterness, inadaptability, impotence, and an inexhaustible longing for escape. As the titles of his first two collections highlight, Casal sought deliverance from the indifferent world that surrounded him in the strange, the foreign (European and oriental), the sick, the dying, and even the unpleasant.

An outstanding example of Casal's struggle with middle-class existence in late-nineteenth-century Cuba is his famous poem "Neurosis," published in *Bustos y rimas* in the year of his death, 1893. The beautiful bourgeois haven filled with imported luxury items is presented with a Parnassian placidity and clarity that deliberately pulls the reader away from the disruptive allusion of both the title and the description of the subject, Noemí, as "la pálida pecadora" ["the pale sinner"].

In his formal analysis, Robert Jay Glickman emphasizes the bipartite structure of the poem. The six-line stanzas are divided into two equal parts of three lines each ("Neurosis" 172–173). The *aab' : ccb'* rhyme scheme underscores the clearly delineated structural division of each stanza. This division, as it appears in the first stanza, anticipates the organization of the rest of the poem. As Glickman points out, the physical description of Noemí in the first three lines serves as an introduction to stanzas 2, 3, and 4, in which her physical environment is portrayed. The suggestive second half of the first stanza alludes to a hidden nature that is more fully developed in stanzas 5, 6, and 7. The first stanza reads:

> Noemí, la pálida pecadora
> de los cabellos color de aurora
> y las pupilas de verde mar,
> entre cojines de raso lila,
> con el espíritu de Dalila,
> deshoja el cáliz de un azahar. (1:231)

[Noemí, the pale sinner / of hair the color of sunrise / and sea-green eyes, / strips the leaves from the calyx of an orange blossom / with a Delilah-like spirit / among lilac satin cushions.]

The young woman, who defiles the purity of nature by destroying the orange blossom, lives within a world of lilac satin cushions. This is the world carefully re-created with Parnassian balance and objectivity in stanzas 2 through 4. Noemí's surroundings are filled with icons of luxury and success. The numerous items range from the burning fireplace and piano and red silk Chinese folding screen to a curved lacquer table and lamp, from a white fan and blue parasol and kidskin gloves to porcelain teacups. These are hard-won trophies of achievement and upward mobility, signaling the triumph of bourgeois values and respectability as clearly as the rigorous structure of the poem signals the triumph of artistic perfectibility over life.

As suggested by the opening stanza, however, the trappings of bourgeois success create a world of appearances that conceals the deeper truths underlying everyday life. Noemí's golden hair and green eyes hide a destructive, Delilah-like spirit. The poem subtly turns to this hidden reality in the last line of the fourth stanza and more directly in the first line of the fifth. Line 24, with its reference to "el alma verde del té" ["the green soul of the tea"], points to a previously unexamined dimension of the scene presented. As the soul of the tea gently escapes from the porcelain teacup, so the emotional and psychological aspects of the scene begin to emerge. Yet they are never as clear as the physical portrayal. They are alluded to through a series of unanswered questions that underscore the fundamental unknowability of psychic states, motives, and desires. They can only hint at the life behind the appearances.

The second half of the poem begins with a "but" that further contrasts the physical and psychological. Casal asks:

Pero ¿qué piensa la hermosa dama?
¿Es que su príncipe ya no la ama
como en los días de amor feliz,
o que en los cofres del gabinete
ya no conserva ningún billete
de los que obtuvo por un desliz? (1:232)

[But what does the beautiful lady think? / Is it that her prince no longer loves her / like in the days of happy love, / or that in the trunks of the parlor / she no longer retains a single note / from the ones she obtained by allowing herself an indiscretion?]

Is she thinking about a lost love or lost funds? Does she reflect on love or passion or commerce? As Noemí sits within the confines of her properly appointed apartment, her stillness hides a multitude of tumultuous possibilities. All those suggested by stanza 5 imply a rebellion against the rules of etiquette and morality assumed to reign within the elegant and orderly interior presented in stanzas 2, 3, and 4. The rules of behavior limit her action and render her immobile. This exterior calm and inactivity is, as clearly stated in the sixth and seventh stanzas, quietly killing her both physically and spiritually. She is conquered by a cruel anemia that leaves her frail and weak. In stanza 6 she is left to reveries understood and perhaps shared by the poet:

> ¿Es que la rinde cruel anemia?
> ¿Es que en sus búcaros de Bohemia
> rayos de luna quiere encerrar,
> o que, con suave mano de seda,
> del blanco cisne que amaba Leda
> ansía las plumas acariciar? (1:232)

[Is it that a cruel anemia overcomes her? / Is it that in Bohemian ceramics / she wishes to enclose moonbeams, / or that, with her soft silken hand, / she longs to caress the feathers / of the white swan that loved Leda?]

Through these questions her illness is tied to issues of transcendence in the most literal of senses. The stanza links her anemia with desires to capture moonbeams in imported ceramic vases or with longings to caress the white feathers of the swan that seduced Leda. Her anemia is linked with going beyond ordinary limits, with struggling to hold for a moment those elements that have been excluded by the materialism and morality of her world. Her isolation from this passion and power, this energy and vision, is the source of her decline.

The advice supplied by "un sueño antiguo" ["an ancient dream"] is "beber en copa de ónix labrado / la roja sangre de un tigre real" ["to drink the red blood of a royal tiger in a sculpted onyx goblet"]. An ancient dream puts her in touch with what modern life has destroyed by forcing her to hide her true nature behind a facade of respectability. Though the poem deliberately leaves room for speculation with regard to what this true nature may be, the occult forces of the bestial, the wild, and the mystical are resurrected as alternatives to the life-draining circumstances in which she exists.

This tension between the physical and the psychological, the visible and the hidden, the modern and the ancient, underscores the desire to reestablish a natural balance that has been destroyed by restrictive conventions and rigid rules—rules imported, for the most part, from Europe and enforced by the ruling classes in bourgeois milieus. In a world of marketable commodities, individual needs together with useful national distinctions are washed away. Superficial success hides the illness of inauthenticity in modern society. From this perspective, the title of the poem aptly highlights the pathology of everyday existence. It reflects, furthermore, an uncanny affinity with Freudian ideas that were just in the process of being formulated and had not yet gained widespread acceptance. Casal gives poetic form to Freud's innovative belief that the origins of neuroses are to be found in sexual fantasies. But perhaps even more important than this extraordinary coincidence is what this perspective suggests about the basis for knowledge. What Peter Gay writes about Freud applies equally to Casal: "Again, as Freud pointed out in his first, still most astonishing, masterpiece, *The Interpretation of Dreams,* in arguing that dreams have meanings that can be understood and interpreted, he was taking the side of the unlettered and the superstitious against blind philosophers and obtuse psychologists" (xvi).

By exploring the dark forces of our inner life, both psychoanalyst and poet renounce the limited view of knowledge established within the materialistic and positivistic epistemological frameworks of the nineteenth century. Despite his refusal to be buttonholed by "scientific" dogma, Freud constantly asserts his membership in the scientific community. Similarly, Casal declares the superiority of his perceptions while grappling with economic and social forces that marginalize him and diminish his stature. Both intellectuals struggle to expand the realm of the explorable, the knowable, and the sayable that has been restricted by dominant assumptions, beliefs, practices, and mores.

These dominant modes of belief and behavior shaped the individual throughout the nineteenth century and became the underlying antagonists within what Frederic Jameson calls "the great modernist thematics of alienation, anomie, solitude, social fragmentation, and isolation" (11). In his *Postmodernism, or The Cultural Logic of Late Capitalism,* Jameson analyzes these factors, focusing on Edward Munch's *The Scream,* a painting that was produced in 1893, the same year as Casal's "Neurosis." Jameson reads *The Scream* as an embodiment not merely of an expression of anguish and anxiety but "as a virtual deconstruction of the very aesthetic of expression itself" (11). What he says about the concept of expression is particularly relevant to Casal's poem, for the very concept of expression presupposes some separation within the subject, "and along with that a whole metaphysics of the inside and outside, of the wordless pain within the [bourgeois] monad and the moment in which, often cathartically, that 'emotion' is then projected out and externalized, as gesture or cry, as desperate communication and the outward dramatization of inward feeling" (11–12). In a similar fashion, Casal's poem points to the need for this type of cathartic experience to release the pent-up pain within the subject. Yet the poem is more tentative than the painting, for it puts that experience within the realm of the private—either within a dream or behind closed doors—and in the ingestion of a prohibited, empowering substance.

From the perspective of the end of the twentieth century, Casal's "Neurosis" sheds light on the ideology of nineteenth-century Cuba, about its power to define and structure the self, and about the artistic response to the confines of those structures. Perhaps even more significant is the degree of coincidence that exists between these features and those that appear in European art and thought of the same period; they reveal *modernismo*'s place within the enormous, international socioeconomic and cultural trends that are shaped by the forces of modernity.

As different as Casal is from Martí, it is their powerful responses to their immediate circumstances that tie them together and make them both *modernistas.* Each in his own way struggles to come to terms with the changing Spanish American scene and each finds in poetry—creative, free, responsive poetry—an outlet for his alternative perspective on his sociopolitical context. Martí's poetry is more spontaneous, Casal's more studied and refined, with a strong emphasis upon pictorial splendor, verbal elegance, and formal innovation—greater accentual flexibility in hendecasyllables, imaginative recourse to nine- and ten-syllable lines, mastery of the monorhyme tercet. Nevertheless the efforts of both poets reflect conscientious attempts to find a language with which to shape the Spanish America that was about to enter the twentieth century.

Like Casal's, José Asunción Silva's life was colored by an attraction to elegance and indulgence and by a series of tragic events, which culminated with his suicide in 1896.[12] Silva grew up in a Bogotá dominated by a conservative and provincial outlook. His family, in contrast, was known for its interest in and support of the arts, as well as for its enthusiasm for imported styles, luxury items, and cosmopolitan trends. Silva's father wrote *artículos de costumbres* [articles on local customs], and the family store was recognized as a center of literary activity. The store, which carried the latest European fashions—even though many of these styles were in little demand in Bogotá—was plagued by financial instability and eventually led the family to bankruptcy. Silva's despondency over these economic woes, which forced him out of school and into the workplace, was compounded by personal losses. His grandfather was killed in a violent attack on the family ranch the year before he was born. His great uncle, who had moved to Paris after the attack, died shortly before Silva arrived in Europe. Even greater strains were produced by the death of his father in 1887 and, in 1891, the death of his beloved sister Elvira, to whom his famous "Nocturno" ["Nocturne"] is dedicated.

After resolving the bankruptcy of the family business, Silva sought a diplomatic post as secretary to the Colombian legation in Caracas. Though his literary efforts there were received with acclaim, he met political difficulties and returned home a few months later. The ship on which he sailed was damaged on reaching the shores of Colombia and began to sink. Though he survived, he lost the bulk of his unpublished manuscripts that had been written earlier in his life and during his stay in Venezuela. This misfortune provided one more devastating blow to the poet, who already felt that life was filled with disappointments and defeats.

Silva had occasionally recited his poems in public and had published a few of them, but most of his work was not widely known in Colombia during his lifetime. Twelve years after his death, a small number of his published pieces and some fifty other poems were gathered together in a volume entitled *Poesías* [*Poems*] (1908), to which Miguel de Unamuno wrote an enthusiastic prologue. Since then at least fifty-three additional poems have been found to belong to the Silva oeuvre. Until recently, his poetry has been divided into three groupings: *El libro de versos* [*Book of Verse*], containing poems from 1891 to 1896 and organized by Silva himself; *Gotas amargas* [*Bitter Drops*]; and *Versos varios* [*Selected Poetry*]. A fourth grouping, *Intimidades* [*Intimacies*], was later added.[13]

In addition to his own poetry, Silva wrote "artistic transpositions" of the works of a number of prominent poets, including Hugo, Gautier, De Guérin, and Tennyson. He also wrote one novel, *De sobremesa* [*After Dinner*]. *De sobremesa* focuses on the longings, tensions, ambiguities, and anguish of a sensitive and involved intellectual, the novel's protagonist, José Fernández, who confronts the developing crises in Spanish America at the turn of the century. This astute portrayal of the critical issues that young artists and thinkers had to face has made *De sobremesa* a cornerstone in the *modernista* movement. Allen Phillips asserts that ". . . Silva's novel is exceptionally important for the study of the intellectual history of *modernismo.* It is, according to our judgment, one of the best documents that we have to know not only the personal crisis of a poet but also the intellectual and international atmosphere of the literary salon, in which the writers of those years moved, at least in their intimate dreams. *De sobremesa* is, then, key and testimony of a whole era" ("Dos protagonistas" 268).[14] I would go even further, however, and propose that it is through this work, with all its quirkiness, that one can come to appreciate the full extent of the imaginative leaps taken by all the early *modernistas. De sobremesa* is a novel that—with its structure, its themes, and its language—underscores the way *modernista* works assert their visionary stance and thereby establish the groundwork for later literary trends.

De sobremesa is a challenging, perplexing, and imaginative narrative that rejects the literary patterns of the end of the nineteenth century. This reluctance to follow predominant novelistic models of the day may explain why early reviews of the novel judged it to be "una obra fallida" ["a frustrated endeavor"].[15] Yet it is precisely with this break with established forms that José Asunción Silva's only novel signals its orientation toward the future and anticipates characteristics more readily linked to works of the end of the twentieth century.

Although many aspects of the novel remain to be studied, critics have repeatedly focused on its enthusiastic embrace of diverse modes of discourse. Silva's work is simultaneously novel and diary, history and fiction, memoir and treatise. Some of the most revealing studies of the narrative have focused on its heteroglossic nature.[16] Evelyn Picon Garfield's fascinating article examines two of the female characters whose real-life existence is played out in the fictionalized world of the poet-protagonist. Benigno Trigo's study examines the presence of end-of-the-century medical discourse throughout the novel. Further fragmenting the nature of the text are its diverse references. Alfredo Villanueva-Collado, in his study of

"La funesta Helena," examines the interweaving of "literary, pictorial, and biographical elements" (67).

While the text contains many starts and stops and relies upon many modes of discourse in addressing its concerns, its overwhelmingly consistent characteristic is its tone. The novel begins with a scene that is prototypically *modernista.*

> Enclosed by the shade of gauze and lace, the warm light of the lamp fell in a circle on the scarlet velvet of the tablecloth; and upon fully illuminating three China cups, golden at the bottom with the remains of thick coffee, and a flask of cut crystal, full of transparent liquor in which particles of gold sparkled, the light left the rest of the silent living drowned in a shadow of dark purple produced by the colorings of the carpet, the tapestries, and the hangings.
>
> Toward the back, attenuated by small shades of reddish gauze, the light from the piano's candles fought with the surrounding semidarkness, while on the open keyboard the ivory played its brilliant whiteness against the ebony's flat blackness.
>
> On the red of the wall, covered by an opaque woolen tapestry, the carvings of the hilts and the polished steel blades of the two crossed swords glittered above a shield; and standing out from the dark background of the canvas, bordered by the gold of a Florentine frame, there smiled, with a good-natured expression, the head of a Flemish burgomaster, copied from Rembrandt. (31) [17]

The elaborate description of a room richly decorated with international artifacts recalls the parlor in which Casal's Noemí sits hoping for release from the confines of bourgeois existence. This initial passage also reveals the novel's affinity to decadentism and includes the most obvious decadent features: self-indulgence, excess, the embrace of artificiality, and a consistent emphasis on both materialism and materiality.[18] The play of light and dark and the shadowy patterns that appear on the tablecloth, the carpet, and the walls, however, point to other features not generally associated with decadentism. These are features that define the entire *modernista* movement and that resurface, distilled into their most pure essence, in two later sonnets by Rubén Darío, who, of course, was confronting many of the same issues at approximately the same time. The intensity of Darío's poetic presentation highlights what tends to escape in the reading of Silva's prose passages. Darío's poems assert the presence of the realities behind the appearance, the meanings behind the surfaces.

The two sonnets are "A Amado Nervo" ["To Amado Nervo"] (446), written in July 1900, and "En las constelaciones" ["In the Constellations"] (*Poesía* 449), written in April 1908. Both form part of Darío's uncollected poetry. "A Amado Nervo" reads:

> La tortuga de oro camina por la alfombra
> y traza por la alfombra un misterioso estigma;
> sobre su carapacho hay grabado un enigma
> y un círculo enigmático se dibuja en su sombra.
> Esos signos nos dicen al Dios que no se nombra
> y ponen en nosotros su autoritario estigma:
> ese círculo encierra la clave del enigma
> que a Minotauro mata y a la Medusa asombra.
> Ramo de sueños, mazo de ideas florecidas
> en explosión de cantos y en floración de vidas,
> sois mi pecho suave, mi pensamiento parco.
> Y cuando hayan pasado las sedas de la fiesta,
> decidme los sutiles efluvios de la orquesta
> y lo que está suspenso entre el violín y el arco.
>
> [A golden tortoise crawls across the carpet,
> and traces on that carpet a mystical stigma;
> its carapace is marked with an enigma,
> and an enigmatic circle is drawn in its shadow.
> Those symbols speak to us of the nameless God
> and mark us with his authoritative stigma;
> that circle contains the key to the enigma
> that slays the Minotaur and dismays Medusa.
> Bouquet of dreams, bouquet of ideas that blossomed
> in an explosion of songs and a flowering of lives,
> you are my gentle breast, my sober thoughts.
> And when the silks of the fiesta are gone, tell me
> of the subtle effluvia of the orchestra, and of what
> is suspended between the violin and the bow.
>
> (Translation by Lysander Kemp 129)]

The similarities between the sonnet and the novel's initial passage are impressive. Both allude to symbolic patterns, economic affluence, and the centrality of music. The poem, however, sheds "light" on what remains "obscure" in the prose text. The sonnet asserts that the pleasures of the flesh must not be allowed to obfuscate the deeper meaning that can be culled from surrounding luxuries. The poet's goal is to convert the wealth of sensations into a magical wisdom that transcends the physical, the material. This magical wisdom is, of course, linked to "el Dios que no se nombra" ["the God that is not named"]. God remains unnamed not only because he cannot be perceived by the uninitiated but also because his identity is fundamentally incompatible with any physical representation, either in image or in word. Yet the poet who aspires to this wisdom is emphatically of this world. His "pecho" is soft and his thought sparing. This assessment of the powers of the poet is not unlike the portrayal of Silva's protagonist, José Fernández, who fails in his (and *modernismo*'s) two grandiose projects: he is unable to execute his political program or to establish a transcendental love relationship.[19]

The burden of trying to reach beyond one's earthbound limitations is even more evident in the second of the two sonnets written by Darío. By the time he writes the second poem, eight years later, after having achieved professional if not material success, Darío confesses: "Pero ¿qué voy a hacer, si estoy atado al potro / en que, ganado el premio, siempre quiero ser otro, / y en que, dos en mí mismo, triunfa uno de los dos? . . ." ["but what am I going to do, if I am tied to the rack / in that, the prize having been won, I always want to be another, / and because, two in myself, one of the two wins? . . ."]. The answer to his dilemma remains the patterns in the sand: "En la arena me enseña la tortuga de oro / hacia dónde conduce de las musas el coro / y en dónde triunfa augusta la voluntad de Dios." ["In the sand the tortoise of gold shows me / toward where the chorus of the muses leads / and where the august will of God triumphs."]

This succinct presentation of the tension between the materiality of existence and the aspiration toward a transcendental plane can be found throughout Silva's poetry as well. In his novel, the opposition emerges through a diverse series of episodes. Whatever may be lost in terms of dramatic intensity by this gradual process is regained through careful elaboration. Played out throughout the novel is a profound dissatisfaction with life in late-nineteenth-century Latin America.[20] José Fernández comes to represent the decadent intellectual who sees society in a downward spiral. As Aníbal González has demonstrated, the novel examines the intellectual's struggle to reach beyond disillusionment and to establish for literature a new authority with which to address the innumerable issues at hand. González explains: "As is clear, the fundamental problematic of literary decadentism was the question of the limits, the boundaries, of literature: the numerous 'isms' in the plastic and literary arts at the end of the nineteenth century and beginning of the twentieth are symptomatic of the anxiety that artists felt in defining and delimiting the nature of their task" (*La novela modernista* 85).[21] He goes on to note: "That meditation about intellectuals in *De sobremesa* centers on the need to define the borders between the world of ideas—of texts, of fictions—that the intellectual deals with and the world of concrete actions" (*La novela modernista* 87).[22] Though Fernández continually goes beyond established boundaries, for González, his "tragedy" is due to his inability to find "a transcendental framework, an absolute limit that would serve him as a point of reference and that would justify the other frameworks" (*La novela modernista* 94).[23] His inability, according to González, turns Fernández into the very "image of the turn-of-the-century writer as a being who deliberately opts to dedicate himself to the production of fictions and to the idolatrous contemplation of them" (*La novela modernista* 112).[24]

This disparaging portrayal of the artist as an individual who prefers to disengage himself from the larger community, however, belies a more encompassing critique that includes epistemological and political concerns. The materiality that appears to lock Fernández within a realm of impotence—either by choice or by ineptitude—is the same materiality with which Darío struggles in the two sonnets quoted above. While individually Fernández—whose common name highlights his inability to escape his Everyman mentality—fails to achieve the goals set out within the text, the entire novel—from the first paragraphs forward—is a constant attack against those structures, frameworks, and conventions that restrict the artist-intellectual.

For Sonya Ingwersen, Gioconda Marún, Howard Fraser, and Alfredo Villanueva-Collado, the greatest departure from traditional perspectives is the recourse to alchemical concepts and imagery throughout the novel. This incorporation underscores the basic impetus toward rediscovering the spiritual wealth that eludes those who live only for the material aspects of life (Fraser, *In the Presence* 95). More specifically, Villanueva-Collado sets out to show that *De sobremesa* is "a hermetic novel that, through alchemical and Rosicrucian symbols, describes a process of spiritual purification to which the protagonist, José Fernández, submits himself under the influence of his love for Helena D'Scilly, who, together with his grandmother, represents the regenerative feminine principle in the novel" ("*De sobremesa*" 10).[25]

Even more important than the details of Villanueva-Collado's analysis is his recognition that Silva's examination of occultist doctrine is in itself a response to the problems associated with modernity ("*De sobremesa*" 20). The transformation of traditional patterns of behavior, the ever greater emphasis on materialism, pragmatism, and utilitarianism, and the insensitivity of the growing bourgeoisie led Silva and many other writers of his generation to explore alternative ways of envisioning the world. Life no longer fit into the comfortable frameworks handed down from the past, and social changes led to a "vulgarization of culture." As a result, a select few dissatisfied intellectuals began to question and challenge socially "acceptable" models of thought and behavior. The philosopher that became an icon of reevaluation and rejection was Nietzsche, and it is not by chance that he appears as such within the novel.[26]

In the section referring to Nietzsche, which—quite significantly—is bracketed by entries of a very different nature, Silva presents, once again, a torn and ambivalent José Fernández (208–212). The entry begins with a statement of indignation, an apparent attack against efforts to overturn current conditions and accepted moral stances. Silva writes:

> Yesterday another building exploded, destroyed by a bomb, and the worldly gathering in a theater on the boulevard applauded, until their hands hurt, *A Doll's House* by Ibsen, a play in the new style, in which the heroine, Nora, a common and ordinary little woman with a common soul, abandons her husband, children, and relations in order to go off to fulfill her obligations to herself, to an "I" that does not know and that feels itself being born one night like a mushroom that appears and grows within a short period of time. So, by explosions of melinite in the bases of the palaces and by spade blows to the depths of its moral foundations, which were the old beliefs, humanity marches toward the ideal realm of justice that Renan believed he could glimpse at the end of time. (177–178)[27]

If the ironic tone of these first sentences were not enough to underscore Fernández's contempt for these revolutionary actions and ideas, he clarifies in the final sentence of the same paragraph his rejection of those intellectuals who seek to restructure traditional patterns of behavior and thought through mass-based efforts. "Ibsen and Ravachol help [humanity], each in his own way. The first magistrate of France falls wounded by Cesáreo Santo's dagger, and Suderman writes *The Woman Dressed in Gray,* where self-denial and family love take on shades of grotesque sentiments. The fairy-tale ending, added by the novelist to his work, like a skillful pharmacist would add a syrup to sweeten a potion that contained strychnine, fails to hide the bitter taste of the lethal drug" (178).[28]

The imagery could not be clearer. Radical change achieved through grass-roots mobilization is deadly. Instead of hearing the call for "fraternity" and "pity for human suffering" proclaimed by Hugo, Dostoyevsky, and Tolstoy, Fernández only hears "the terrible voice of Nietzsche." In a lengthy section, Fernández appears persuaded by Nietzsche's rejection of dominant belief structures but disgusted by the workers whom he believes would be the greatest beneficiaries of any revolutionary action. A brief passage is sufficient to underscore this fundamental tension. "Hey, worker, you who spend your life doubled over in two, whose muscles are impoverished by rough work and deficient nutrition, but whose calloused hands still make the sign of the cross, worker, you who bend your knee to petition heaven on behalf of the owners of the factory where you are poisoned with the vapors of the explosive mixes, hey, worker, do the rough syllables of that German name, Nietzsche, when they vibrate in your ears, evoke nothing in your rudimentary brain?" (178–179).[29]

Fernández's understanding of the abuse suffered by the worker at the hands of the powerful is contradicted by his insulting and dehumanizing tone. He appears oblivious to the possibility that the presumed ignorance of the workers is produced by the ideologies, both secular and religious, that, under the influence of Nietzsche, Fernández deems to be false. "It is that humanity had been receiving as true false notions about its origin and its destiny, and the profound philosopher found a touchstone upon which to assay ideas like coins are assayed in order to know their gold content. That is what is called revaluing all values" (179).[30]

Fernández continues, in apparent agreement with Nietzsche, a discussion of the origin of beliefs. He states that all belief systems—from individual conscience to Christian morality—are a product of their formation, a product that reflects numerous factors, some of which are psychological and some of which are political but none of which is based on absolute truth. Fernández's adoption of this Nietzschean premise gives him leave to reject outright the confines of traditional morality and to propose the freedom of self-creation associated with the superman, the *Übermensch.* But this freedom is immediately presented as a two-edged sword. The journal entry continues:

If conscience is the claws with which you hurt yourself and with which you can destroy what is presented to you and you can seize your part of the booty of victory, do not sink them into your flesh, turn them outward; be the superman, the *Übermensch* free of all prejudice, and with the calloused hands with which you, stupid, still make the sign of the cross, pick up a little of the explosive mixtures that poison you as you breathe their fumes, and make the ostentatious dwelling of the rich individual that exploits you break into pieces, with the explosion of fulminatory picrate. Once the masters are dead, the slaves will be the owners and they will profess the true morality in which lust, murder, and violence are virtues. Do you understand, worker? . . . (180)[31]

In this vision, Fernández distills the social, political, and philosophic struggles that developed in the nineteenth century and that continued throughout the twentieth. The ironic shift in tone at the end of the last passage quoted once again points to his ambivalence—an ambivalence that, needless to say, runs throughout *modernista* texts and that is reflected in a rejection of and longing for inclusion among the privileged. At first Fernández seems to present a Marxist position, in which he reveals his disgust at the exploitation of workers for the benefit of a few powerful owners. He immediately follows this sympathetic rendering of the workers with Nietzschean disdain for the masses and all herd mentalities. He is disgusted by the vision of those ignorant individuals who would misuse the possibilities provided by the Nietzschean undermining of truth and values and the exaltation of the will to power. Yet clearly there are those—and Fernández includes himself among them—that would use this freedom well. Fernández's entire megalomaniacal plan for ruling his homeland grows out of this Nietzschean perspective. Though terrified by the thought of the uncouth and uninformed asserting their collective will to redefine morality and behavior, Fernández feels entitled to break all established rules of morality and behavior.

By indicating the origin of this sense of superiority, this journal entry clarifies much of what Silva does in the novel. Silva shows himself sympathetic to those radical views about truth and reality that have led many to identify Nietzsche as the grandfather of postmodern thought (and with this reference I am deliberately underscoring an early link with later literary developments). Silva's rejection of any one mode of discourse, his creative fusing of history and fiction, his assertive defiance of bourgeois sensibilities, all point to a profound understanding of how thinkers, writers, politicians, and individuals have been limited, blinded, unreasonably restricted by preestablished rules, traditional modes of thinking, and falsely constructed philosophic groundings.

Silva even mocks—through Fernández in this very same passage—popularized versions of unorthodox beliefs that the author himself incorporates into his text, beliefs that would offer consolation to many of his generation. Although the search for a transcendental worldview is essential to the novel, Silva does not hesitate to attack the fashionable tendency to adopt ready-made solutions to troubling realities.

> . . . Look: from the dark depth of the Orient, homeland of the gods, Buddhism and magic return to reconquer the Western world. Paris, the metropolis, opens its doors to them like Rome opened them to the cults of Mitra and Isis; there are fifty theosophic centers, hundreds of societies that investigate mysterious psychic phenomena; Tolstoy abandons art in order to make practical propaganda for charity and altruism, humanity is saved, the new faith lights its torches in order to illuminate the dark path! (181–182)[32]

Christianity has been undermined by positivistic thought and by philosophic reassessments of truth and values. Alternatives have been sought, but Fernández astutely recognizes the corrupting impact of readapting these alternatives for mass-consumption. The search for truth has become a quasicommercial endeavor, with hundreds of "centers" cropping up to fill the spiritual void of the moment. With this commentary, Silva shows a remarkable perceptivity about his own historical context. He seems to have had his fingers on the pulse of society, cognizant of every struggle for power, every philosophic realignment. Projecting this awareness on Fernández does not, however, free him from his human foibles and drives. Quite the contrary, Silva chooses to highlight just how inextricably bound Fernández is by his humanity. Even the framing of this important journal entry indicates the recognition that, while people may speculate freely, they are often restrained by passion, desire, or emotion.

The entries before and after the section dated April 14, which has just been examined, focus on Fernández's love for Helena. The one dated April 13 ends with a passionate call: "Helena! Helena! Fire runs through my veins and my soul forgets about the earth when I think about those hours that will arrive if I succeed in finding you and uniting your life with mine! . . ." (177).[33] The one dated April 15 begins: "A powerful wave of sensualism runs through my whole body, ignites my blood, strengthens my muscles . . ." (183).[34] As a result the reader moves from a whirlwind of passion to the heights of philosophic discourse back to a whirlwind of sensuality. In this tug to and from the "beastlike" impulses of human existence, Fernández offers a parallel to Nietzsche's tightrope walker in *Thus Spake Zarathustra.*

At one point in this seminal work, Nietzsche has Zarathustra proclaim: "Man is a rope, tied between beast and overman—a rope over an abyss. A dangerous across, a dangerous on-the-way, a dangerous looking-back, a dangerous shuddering and stopping" (126). While this passage reflects long-established views of creation often identified with the concept of the great chain of being—views that were modernized by Darwinian concepts known to Nietzsche—it also captures Fernández's great challenge as well as the one Darío addressed in the two sonnets quoted earlier. Nietzsche's tightrope walker must move from one extreme to the other. When he falls as he tries to cross above the marketplace and fails to reach his final destination, Zarathrustra offers the following consolation: "You have made danger your vocation; there is nothing contemptible in that. Now you perish of your vocation: for that I will bury you with my own hands" (132).

Though Silva portrays Fernández—like the poetic voice of Darío's two sonnets—as pulled toward the beastlike, he opens, through the novel, possibilities that Fernández is unwilling to pursue. The novel, criticized as flawed, broke rules in its search for the freedom to see beyond the restrictive materialism and political self-interest of the day. Even if we judge Fernández to have failed, we are now beginning to recognize—from our postmodern point of view—that the novel did not. It took up the challenge of making danger its vocation. It sought to address issues of political and spiritual transcendence at the expense of personal comfort and conformity. It underscores the arduous task of trying to carry out *modernismo*'s dual mission.

As in *De sobremesa,* Silva's poetry refused to follow easy models. His most acclaimed poem, "Una noche . . ." ["One night . . ."] (131–133), generally called "Nocturno," is often quoted as one of *modernismo*'s boldest attempts to alter the structures of Spanish poetry. Yet it did much more. Drawing upon Silva's creative use of the rhythmic resonances of the Spanish language—enhanced by a free use of repetition—it offers a vision in which nature is eternal, divine, and harmonious. At the same time, it presents the symbolic evocation of subtle psychological and spiritual states that respect the limits of cognitive comprehension and the impermeability of certain mysteries. "Una noche . . .", through its interplay of light and shadow, which is one with its central and structuring metaphor of "la sombra nupcial" ["a nuptial shadow"], suggests a reality that evades precise definition.

For this reason, when *De sobremesa*'s José Fernández discusses his own writing and declares that he did not want his poetry to state but rather to suggest, most critics have assumed that Silva was speaking for himself. More than any other poet of his generation, Silva wrote under the influence of symbolist poetics with its goal of transforming the music of nature into an evocative language that captures the eternal and profound realities that lie hidden behind the surface of existence. This aspiration—together with a political indictment against his unworthy times—is expressed in "Al pie de la estatua" ["At the Foot of the Statue"] (233–244), the great epic and civil poem about Bolívar written in and for Caracas. Finished just six months before his death, this text captures a sense of disillusionment at the same time that it expresses the concepts that Darío, in the same year, placed at the center of his "Coloquio de los centauros" ["Colloquy among the Centaurs"] and that Baudelaire had outlined in his famous "Correspondances." Nature speaks to the poet, whose task is to translate into poetry its cosmic signs and symbols and its hidden and vital forces.

In poems such as "Vejeces" ["Old Things"] (137–138) or "La voz de las cosas" ["The Voice of Things"] (134), the timelessness of the spirit is contrasted with the fugacity of individual existence. Despite the solace thus evoked, there remains in many poems a sense of suffering and loss reflective of Silva's tendency to indulge in a mournful and morose analysis of unhappiness. In some pieces from *Gotas amargas,* a collection that Silva seems to have planned to leave unpublished, suffering results from the incomprehension of the scientific community, which is mocked in acerbic and sarcastic tones. The resulting satire highlights the pain, anguish, and anger of the *modernista* struggle with "bourgeois modernity" and with the dominant positivistic values of the day. With his sights set on the exploration of the mysteries that elude modern science, Silva had little patience for Parnassian aesthetic play or popular but superficial imitations of Darío's work, a point well made in his satiric "Sinfonía color de fresas con leche" ["Symphony the Color of Strawberries and Milk"] (227–228). As María Mercedes Carranza recognizes, even though these fifteen poems are generally judged to be of lesser importance, here again Silva proves himself to be an unexpected precursor, this time of the antipoetry of Luis Carlos López and Nicanor Parra (25).

Silva's work, his prose as well as his poetry, is permeated by a perception that life would always fall short of his expectations and hopes. Yet for as long as he lived, he struggled to see beyond the ordinary and to propose the extraordinary—for the arts, for his community, and for his own sense of personal consolation. Yet it was his critique of the social changes that were occurring and of the intellectual and artistic responses to them that turned Silva into a visionary. He was out of step with, but often ahead of, his times. This attitude he shared with Rubén Darío, the focus of the next chapter.

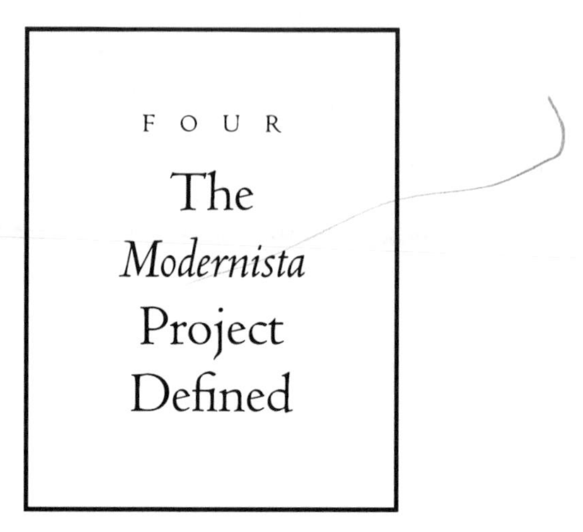

FOUR
The *Modernista* Project Defined

FOR many, the name Rubén Darío is synonymous with *modernismo.* His artistic production spans the entire movement, and his talent has been favorably compared with that of the greatest poets of the Spanish language. Darío's artistry so completely eclipsed the other *modernistas* that, in 1967, Enrique Anderson Imbert felt free to write: "The *modernismo* that interests me now—I do not deny that there may be others—is the one that is born and dies with Rubén Darío" (*La originalidad* 17).[1] Similarly, several generations of critics so closely identified the *modernista* movement with Darío that they denied the four poets just discussed full credit for their artistic innovations and influential breakthroughs, calling them "precursors." While scholars have now set aside this misconception, Darío is still considered the central figure of the movement. His centrality, while tied to his extraordinary talent, derives from two key aspects that are the focus of this chapter. His poetry elegantly captures the essence of the dominant intellectual and artistic issues of the day. His life and works, moreover, reflect the trends and tendencies that reveal the complex relationship between the development of *modernismo* and the evolution of modern life in Spanish America.

He was born Félix Rubén García Sarmiento on January 18, 1867, in a small town renamed Ciudad Darío in his honor. The Darío by which he was known all his life was not so much a pseudonym as a patronymic, the last name used by his father as well as his grandfather. Shortly after his birth, his parents separated and he was taken to live with his great aunt and her husband in León. Thus began his peripatetic life, which took him to all corners of the Hispanic world, driven for the most part by economic necessity. By the age of twelve he had already published his first poems. During these early years, there were frequent trips among the cities of Managua, Granada, and León. In 1882 he lived for a year in El Salvador. There the poet Francisco Gavidia introduced Darío to French literature—to the formal beauty of Parnassian verse, to the Gnostic and pantheistic writings of Victor Hugo, and to the alexandrine and hexameter. By 1886, when he left Central America for Chile, Darío had read the Spanish classics, Greek poetry in translation, and contemporary French verse published in the *Revue des Deux Mondes.*

His three years in Chile, from 1886 to 1889, offered the young Darío experiences in Santiago, Valparaiso, and other rapidly changing parts of Spanish America. His stay there also provided an expanded arena of cultural explorations: a cosmopolitan setting, friendship with many writers and intellectuals, and contact with a society that, because of its growing prosperity, enjoyed sophisticated manners of behavior, dress, and patronage. During this period, Darío wrote for important newspapers like Santiago's *La Época,* where he developed an elegant and effective prose style. At the same time, he published his first two books of poetry, *Abrojos* [*Difficulties*] (1887) and *Otoñales* (*Rimas*) [*Autumnal Poems* (*Rhymes*)] (1887). He also wrote the novel *Emelina* [*Emeline*] and the award-winning *Canto épico a las glorias de Chile* [*Epic Song to the Glories of Chile*]. Even in these early pieces there are hints of the metrical experimentation, the emotional depth, and the metaphorical brilliance that would characterize his mature production. Yet stronger at this point was the influence of his immediate predecessors, most notably Bécquer and Campoamor.

Azul . . . [*Blue . . .*], first published in 1888, was the volume that made Darío famous and by which critics used to date the beginning of *modernismo.* The already cited letter by Juan Valera turned Darío and his work into the focus of attention for those interested in the revolutionary but still unnamed literary movement that was taking shape in Spanish America. It is widely agreed that *Azul . . .* 's short stories and vignettes, written under the influence of Mendès, Flaubert, Hugo, and Zola, were more formally innovative than its poetry. The entire collection, however, breaks ground and unveils certain characteristics that prove fundamental to the evolving movement. The prose pieces show Darío's eagerness to experiment with many styles and modes of discourse, his enthusiasm for and emulation of the various arts, and his desire to break, as the romantics had done, the restrictive confines of established genres. In the poetry, especially in its recourse to overtly sexual metaphors, Darío further asserts his intent to confront convention and to break accepted norms.[2] These literary acts of noncompliance challenge established patterns of perception and reflect his disillusionment—which he shared with many other Spanish Americans writing at the time—with a society that both elevated the mundane and the pedestrian and tended to ignore the aesthetic and the spiritual. "El rubí" ["The Ruby"], "El sátiro sordo" ["The Deaf Satyr"], "El palacio del sol" ["The Palace of the Sun"], "El rey burgués" ["The Bourgeois King"], and, perhaps most directly, the introductory section of "En Chile" ["In Chile"] (all from *Azul . . .*) criticize the limited and limiting vision of bourgeois materialism, science, and technology. "En Chile" begins with a paragraph-long sentence that reveals the focus of Darío's writings at this point and recalls Gutiérrez Nájera's desire to distance himself from the chaos and confusion of modern life.

> Without brushes, without palette, without paper, without pencil, fleeing the excitement and confusion, the machines and bundles, the monotonous noise of the trolleys and the jostling of horses with their ringing of hooves on the stones, the throng of merchants, the shouts of newspaper vendors, the incessant bustle and unending fervor of this port in search of impressions and scenes, Ricardo, an incorrigible lyric poet, climbed up Happy Hill, which, elegant like a great flowering rock, displays its green sides, its mound crowned by smiling houses terraced at the summit, homes surrounded by gardens, with waving curtains of vines, cages of birds, vases of flowers, attractive railings, and blond children with angelic faces. (*Azul. Prosas profanas* 113)[3]

The world of the modern industrial city with its traffic, noise, and newspapers (the commercial side of writing) is left behind in search of "impressions and scenes," that is, in search of a nature filtered through, captured in (like the caged birds and cut flowers), and idealized by art. The desperation that drove Gutiérrez Nájera to explore suicide as an attractive option in "Para entonces" is immediately brought under control. Despair is assuaged by the redemptive power of art. By the time this faith reappears in *Cantos de vida y esperanza* [*Songs of Life and Hope*], seventeen years later, it has become an entrenched dictum of *modernismo:* "y si hubo áspera hiel en mi existencia, / melificó toda acritud el Arte" (*Poesía* 245) ["anc was sour bile in my existence, / Art turned all bitterness into hon

In "En Chile," Ricardo leaves the Valparaiso "that performs tran and that walks like a gust, that peoples the stores and invades the banks" in hopes of finding "el inmenso espacio azul" ["the immense blue space"] (*Azul. Prosas profanas* 113). He seeks not only the free, clear sky of peace and tranquility but also the source of artistic inspiration which transforms the author into a seer capable of discovering the profound realities of existence, an existence that is in essence beauty and harmony and not the crass world of marketable commodities that permeates the urban landscape. This is the point Darío alludes to in his title, *Azul . . .*, which recalls numerous works of the period and evokes the ideal, the infinite, and the eternal—all perceived through and re-created in art.[4]

The poems of the first edition of *Azul . . .* reflect the same tensions and longings as the prose, but they inject the element of erotic passion. The four-part "El año lírico" ["The Lyric Year"] (*Poesía* 157–170), which begins the poetic selection, offers an escape from the mundane similar to that found in "En Chile." Both pieces propose an alternative to immediate circumstances. The poetic grouping, however, finds its ideal context in exotic, fanciful, and exquisite settings. More significantly, the poems link the fundamental aspiration toward harmony to the fulfillment of sexual desire—unencumbered by social expectations. Woman is more than the poet's Muse. She is the other that complements and completes, the one with whom the poet attains a vision of beauty, harmony, and artistic perfection that is simultaneously in tune with and supported by nature. The chasm between this dreamed realm of perfection and contemporary "civilized" existence is brought out in "Estival" ["Of Summer"] (*Poesía* 160–164). In this, the second poem of the grouping, the flow of sexual energy, which is portrayed as the animating force in nature and the inexorable bond between male and female, is disrupted by a cruel and senseless act on the part of the Prince of Wales. Power and modern technology burst upon a scene of lush sensuality and animalistic eroticism, interrupting the natural order of things. "Estival," perhaps naively, thus emphasizes how human intervention by the unenlightened destabilizes the balance of creation and unleashes violence, pain, and discord.[5]

This affirmation of sexuality becomes a centerpiece of Darío's poetic worldview. It appears not only in *Azul . . .* but also throughout Darío's lifework, indeed throughout *modernista* writings in general. Scholars have recognized in this eroticism a type of mysticism, but most have ignored the link between overt sexuality and the rejection of social constraints or religious formulas that failed to provide a satisfactory sense of order and tranquility.[6] Instead of finding in middle-class patterns of behavior welcome guidelines or comforting routines, Darío and other *modernistas* saw distortive restraints that distanced the subject from natural tendencies that they considered to be eternal, liberating, and illuminating. In contrast, these writers came to regard the erotic embrace of lovers as one of the harmonious rhythms of existence, equal to the beating of hearts, the lapping of waves, or the music of the spheres. To avoid sexual contact or to alter its character was to accede to social pressures that they no longer considered worthy of respect. The defiance expressed rejects bourgeois inhibitions and champions spiritual liberation.[7] As already noted, in both the poetry of Casal and Silva's novel *De sobremesa,* sexual decorum became emblematic of social conventions, which serve the privileged while imposing unsatisfactory limitations and false appearances that prevent a truer, more intuitive understanding of life. Though Martí's writing is less emphatically sexual than that of other *modernistas,* he adheres to this same overall vision in the prologue to *El poema del Niágara.* What began with these early writers continued throughout the movement.

In the 1890 edition of *Azul . . . ,* Darío included a number of additional poems that highlight his rapid maturation as a poet. "Venus" takes up the themes of the earlier edition while it advances—with its unusual seventeen-syllable lines—efforts to expand the poetic potential of the Spanish language. The idealization of love seen before is taken to a higher level as the unnamed object of desire is identified with the star-goddess Venus. Here, however, the unreflective hope for ecstasy of the earlier poems is placed in doubt. The expectation of a perfect union between personal and artistic goals is dashed as the heaven from which Venus looks down upon the poet is turned into an abyss of unfulfilled and possibly unfulfillable longings. The tercets tell the story of this transition from hope to despair.

> "¡Oh, reina rubia!,—díjele—, mi alma quiere dejar su crisálida
> y volar hacia ti, y tus labios de fuego besar;
> y flotar en el nimbo que derrama en tu frente luz pálida,
> y en siderales éxtasis no dejarte un momento de amar".
> El aire de la noche refrescaba la atmósfera cálida.
> Venus, desde el abismo, me miraba con triste mirar. (*Poesía* 175)

["Oh, blonde queen!," I said, "my soul wishes to leave its chrysalis / and to fly to you, and to kiss your lips of fire; / and to float in the nimbus that spills pale light on your forehead,

and in starry ecstasy to not stop loving you for a moment." / The night air cooled down the warm atmosphere. / Venus, from the abyss, was looking at me with a sad look.]

Once again Darío, with this early poem, establishes a feature that was to remain fundamental to his entire oeuvre. In "Venus," Darío spotlights the tension between optimism and hopelessness that imbues his writings with an energizing spark and a recognizably modern anxiety. The figure of the modern poet emerges, teetering at the edge of despondency, fully aware of all dangers and pitfalls before him. His responsibility is to shed his earthbound flaws and failings and achieve transcendence through the loving embrace of beauty and freedom. His aspirations are lofty, the possibilities of failure enormous. Yet he believes, at times against all reason, that art, beauty, and devotion—all three as one—may provide a solution, a salvation, an alternative to everyday existence.[8]

The other sonnets that Darío added to the 1890 edition reflect this obsessive preoccupation with poetic aspirations. They deal with historical figures—Caupolicán, Leconte de Lisle, Catulle Mendès, Walt Whitman, J. J. Palma, Salvador Díaz Mirón, and Alessandro Parodi—who are praised for embodying traits consonant with the goals that he envisioned for *modernista* verse: a soul that is in touch with the world and capable of prophetic powers ("Walt Whitman"), a song that echoes the rhythmic pulsation of the ocean and that contains the mysteries of the Orient ("Leconte de Lisle"), classic grace and intimate knowledge ("Catulle Mendès"), and powerful poetry that proclaims the freedom of the new nations of the New World ("Salvador Díaz Mirón"). Taken as a unit, these poems reinforce the epistemological and political nature of the *modernista* project.

Between the first and second editions of *Azul . . .*, Darío faced serious economic difficulties and was forced to return to Nicaragua in February of 1889. After spending a few months there, he left for El Salvador to manage *La Unión*, where he fell in love with Rafaela Contreras and married her on June 26, 1890. Just then a military coup overthrew his protector, President Francisco Menéndez, and Darío was compelled to leave for Guatemala. In Guatemala he became friends with Jorge Castro, who introduced him to theosophy and other occultist beliefs that rounded out the readings that he had begun earlier with Gavidia. While in Guatemala he also published the augmented version of *Azul. . . .* Yet, because of the demise of the newspaper that he was directing, Rafaela and he were once again obliged to move on, this time to Costa Rica, where his son Rubén Darío Contreras was born on November 12, 1891. By May 1892 further economic problems caused Darío to leave his wife and son as he returned to Guatemala alone. There he was named secretary to the Nicaraguan delegation to the Fourth Centenary of the Discovery of America, to be celebrated in Madrid on October 12, 1892.

In Madrid he met members of the old guard of Spanish letters, including Zorrilla, Valera, Castelar, Núñez de Arce, Pardo Bazán, and Campoamor, as well as the Spanish *modernistas* headed by Salvador Rueda, for whose *En tropel* [*Tumultously*] he wrote a verse prologue. While in Spain, Darío extended his fame and consolidated his stature as leader of the *modernista* movement. After a few months he returned to Nicaragua, with a stopover first in Havana so that he could meet Julián del Casal, whose *Hojas al viento* and *Nieve* had explored issues similar to those that Darío was addressing in his own work. Upon his return to his homeland, he learned of his wife's death in El Salvador. Though he always remembered Rafaela as the ideal bride, he was quickly remarried, perhaps with no choice in the matter, to Rosario Emelina Murillo. In April 1893 he was named consul general of Colombia in Buenos Aires. Darío left Rosario in Panama and headed for Buenos Aires by way of New York and Paris. In New York he met Martí, whose work—especially his prose—Darío had long admired. In Paris Darío witnessed the heyday of symbolism, met with Paul Verlaine, and befriended Jean Moréas. He also spent time with Guatemalan-born Enrique Gómez Carrillo, who is remembered today for his travel chronicles, and with Alejandro Sawa, a nearly forgotten Spanish writer known for his Bohemian lifestyle.[9]

Darío arrived in Buenos Aires in August of 1893. Although the position as consul-general did not last long, he was not totally without financial support; he had already been invited to work for the best newspapers in Argentina. During his five years in Buenos Aires, from 1893 to 1898, Darío continued to write both prose and poetry. With Jaimes Freyre he founded the *Revista de América,* and, in 1896, he published an important and revealing series of articles on modern European and American authors entitled *Los raros* [*The Odd Ones*]. About the same time he began a novel called *El hombre de oro* [*The Man of Gold*].

It was, nevertheless, the publication of *Prosas profanas* in 1896 that marked a watershed in the *modernista* movement. Darío saw himself at the head of the dominant Hispanic literary movement of the day, in full control of his talents, and in a position to challenge—with his inventive title, his poetic innovations, and the overt eroticism of most of the poems—conservative critics, unsympathetic members of society, and less creative rivals. His encyclopedic grasp of culture, his syncretic imagination, and his keen sense of direction and purpose allowed him to assume the responsibilities left to him by the death of so many of the other early *modernistas.* His views stated in the nonmanifesto manifesto at the beginning of *Prosas profanas,* "Palabras liminares" (*Poesía* 179–181), and in its three masterly introductory poems constitute a careful, comprehensive, and insightful commentary on *modernismo,* one reflective of the essential elements that shaped the movement since its inception.[10]

Prosas profanas is often described as a youthful, exuberant work full of exotic frivolity, playful imagination, and pleasure. When Darío himself refers to the content of the collection and its title, he directs attention toward sexual passion—a sexual passion that is inextricably linked to art, poetic creation, music, and religion. He wrote: "I have said, in the pink Mass of my youth, my antiphons, my sequences, and my profane proses. . . . Ring, bells of gold, bells of silver; ring every day, calling me to the party in which eyes of fire shine, and the roses of mouths bleed unique delights" (*Poesía* 180).[11] The proses, like the antiphons and sequences, are verses or hymns said or sung during the Mass. Darío plays with these medieval Catholic allusions, breaks expectations regarding the genre in question, and by equating divine love and religious devotion with sexual exploits, defiantly transgresses accepted norms and proposes a broad realignment of values.

However much the young Darío was preoccupied with erotic longings, he was equally fascinated—as noted earlier—by the limits, restrictions, and constraints imposed on behavior, language, and vision by society, the same constraints that dismayed Martí, tortured Casal, and haunted Silva. As a result, the sociocultural context of *modernismo* is never far from Darío's mind. He begins "Palabras liminares" with regret over the lack of understanding common to the general public and to professionals. His criticism is specifically directed against the insensitive and closed-minded middle-class individuals (Rémy de Gourmont's fictional "*Celui-qui-ne-comprend-pas*") that are found in all professions, even the arts. Yet it is art that sets him—and the others who would rally to his cause—apart. Art, however, is not imitation; it is the transgressing of limits; it is the reinterpretation and revitalization of models and rules by each artist, who, as Martí asserted in 1882, must not be constrained by the patterns set by others.

The interplay between poetry and society reappears in the aristocratic and exotic elements that Darío offers in response to the rampant materialism and overwhelming insensitivity around him. This conflict between the crass values of society and the poet's aesthetic and visionary aspirations forms the background to his declaration "I detest the life and times to which I was born." This statement is not, however, a rejection of Spanish America. Quite the contrary, as he did in *Azul . . .* , Darío finds poetry in "our America" in "the old things"—in Palenke and Utatlán, in the sensual and refined Inca, and in the great Montezuma. He yearns for a cosmopolis and future in which the Spanish, the Spanish American, and the European (Parisian) would find a balance. The milieu that he envisions would lead to a modern mode of discourse, that is, a poetry that overcomes the restrictions of modern life by rediscovering the soul of language and its musical nature ("Since each word has a soul, there is in each verse, in addition to verbal harmony, an ideal melody. The music comes exclusively from the idea, many times" [*Poesía* 180][12]).

The reference to the soul of language implies a body which, in Darío's poetry, is clearly female. For language to become poetry it must be inseminated with ideas that are in essence "an ideal melody." This image of poetic creation permeates the views of woman and sexuality expressed throughout *Prosas profanas.* For example, to acknowledge the influences on his work Darío declares: "my wife is from my homeland; my mistress, from Paris." Similarly, he concludes "Palabras liminares" with the mandate: "And the first law, creator: create. Let the eunuch snort. When a Muse gives you a child, let the other eight be left pregnant" (*Poesía* 181).[13] Despite the jocular tone of this command, Darío is never blind to the possibility, alluded to in "Venus," that he may not find the female other that he seeks. This concern continues into the first three poems of *Prosas profanas.*

Darío begins with Eulalia of "Era un aire suave . . ." ["It was a gentle air . . ."]. He characterizes her—or, actually, her golden laughter—as cruel, thereby softening the bold and ambitious declaration of artistic goals of the prose preface. He acknowledges the possible recalcitrance of poetic language to be molded to the form he envisions. By calling Eulalia eternal, he affirms his aspiration—and that of the other *modernistas*—to take Spanish American discourse out of its anachronistic present and to have it become "modern" through a syncretic exaltation of the beauty and art of all ages. For this reason, the second part of "Era un aire suave . . ." is a series of questions regarding the setting of the first part.

¿Fue acaso en el tiempo del rey Luis de Francia,
sol con corte de astros, en campos de azur?
¿Cuando los alcázares llenó de fragancia
la regia y pomposa rosa Pompadour?
¿Fue cuando la bella su falda cogía
con dedos de ninfa, bailando el minué,
y de los compases el ritmo seguía
sobre el tacón rojo, lindo y leve el pie?
¿O cuando pastoras de floridos valles
ornaban con cintas sus albos corderos,
y oían, divinas Tirsis de Versalles,
las declaraciones de sus caballeros?
¿Fue en ese buen tiempo de duques pastores,
de amantes princesas y tiernos galanes,
cuando entre sonrisas y perlas y flores
iban las casacas de los chambelanes?
¿Fue acaso en el Norte o en el Mediodía?
Yo el tiempo y el día y el país ignoro,
pero sé que Eulalia ríe todavía,
¡y es cruel y eterna su risa de oro! (*Poesía* 182–183)

[Was it perhaps in the time of King Louis of France, / sun with a court of stars, on a field of blue? / When the royal and magnificent Pompadour rose / filled the fortresses with fragrance?

Was it when the beauty picked up her skirt / with nymphlike fingers, dancing the minuet, / and, above the red heel, the pretty and light foot followed / the pace of the rhythms?

Or when the shepherdesses from flower-filled valleys / decorated their white sheep with ribbons, / and divine Thyrses of Versailles heard / their gentlemen's declarations?

Was it in those good times of shepherding dukes, / of loving princesses and tender young men, / when among smiles and pearls and flowers / the coats of the chamberlains would pass by?

Was it perhaps in the North or in the South? / I do no know the time or the day or the country, / but I know that Eulalia still laughs, / and her golden laugh is cruel and eternal!]

By proposing alternative periods to which the scene described could belong and by suggesting that there is no need to resolve the riddle ("Yo el tiempo y el día y el país ignoro"), the poet asserts his desire to transcend temporal and spatial limitations and to achieve universality. The human, particularistic elements are clearly subsumed to the creation of a mythic time and place, a "Golden Age" in which art takes over.

These final stanzas provide an explanation for the deliberate ambiguity of the first half of the poem. The uncertainty regarding the poem's setting led Anderson Imbert to ask: "Museum, masked ball, journey through time?" (*La originalidad* 78). "Era un aire suave . . ." is the *modernista* embrace and appropriation of all three. The wealth of images, the luxury of details, and the consolidation of knowledge all belong to the poet, who, at the perfect point in the timeless evening of the poem, will join with Eulalia. Surrounded by auspicious and evocative music and an ivory-white swan of formal beauty and fluid grace, he will vanquish his rivals, the "vizconde rubio" ["the blond viscount"] and the "abate joven" ["the young abbot"]. This reference to the defeat of his social and literary competitors (most notably, Verlaine) as he enters, virtually for the first time, the *fêtes galantes* is a declaration of enormous confidence.[14] The poet's happiness, however, is mitigated by the fact that there is no lasting amorous conquest. On the contrary, he remains her page, her servant, haunted by Eulalia's mocking laughter.[15] With this emphasis on Eulalia's aloof nature and the possible intractability of poetic language, "Era una aire suave . . ." hints at what would become a full-blown lament in "Yo persigo una forma . . ." ["I pursue a form . . ."] (*Poesía* 240–241), which appeared, five years later, as the closing piece of the 1901 edition of *Prosas profanas.* The incorporation of many of the images of "Era un aire suave . . ." in this final poem serves to underscore both the intent of the first poem and the progressive erosion of the poet's self-confidence.

In the meantime, however, Darío's cautiously optimistic response to the proliferation of cultural elements at the end of the nineteenth century that is central to "Era un aire suave . . ." is also the focus of "Divagación" ["Wandering"], the second poem in *Prosas profanas.*[16] "Divagación" is filled with cosmopolitan references, exquisite vocabulary, and esoteric proper names.[17] Like "Era un aire suave . . . ," it deals with a beloved who stands for much more than a possible love interest. Yet throughout his poetic journey across the globe, the poet finds that no one woman can satisfy; no one style can fulfill his longing for an original mode of discourse. The poet's aspiration to an all-encompassing grasp of reality takes him through a literary "museum," which he ultimately leaves behind.[18] He affirms instead the power of poetry, through which he claims divine knowledge and authority. He makes this assertion in the final three stanzas of the poem, in which he leads the reader off the map into the world of the transcendental, thereby emphasizing the divine mission that he strives to achieve.

> Amor, en fin, que todo diga y cante,
> amor que encante y deje sorprendida
> a la serpiente de ojos de diamante
> que está enroscada al árbol de la vida.
> Ámame así, fatal, cosmopolita,
> universal, inmensa, única, sola
> y todas; misteriosa y erudita:
> ámame mar y nube, espuma y ola.
> Sé mi reina de Saba, mi tesoro;
> descansa en mis palacios solitarios.
> Duerme. Yo encenderé los incensarios.
> Y junto a mi unicornio cuerno de oro,
> tendrán rosas y miel tus dromedarios. (*Poesía* 186–187)

[Love, in short, that says and sings all, / love that charms and leaves surprised / the serpent of diamond eyes / that is wrapped around the tree of life.

Love me like this, fatal, cosmopolitan, / universal, immense, unique, one / and all; mysterious and erudite: / love me sea and cloud, foam and wave.

Be my Queen of Sheba, my treasure; / rest in my solitary palaces. / Sleep. I will light the incensories. / And beside my unicorn's golden horn, / your dromedaries will have roses and honey.]

As clearly stated in the first of these three final stanzas, Darío is seeking a love that is, in essence, harmony. Union with this woman would reverse the divisive and disruptive results of the fall. It would provide a prelapsarian vision that could transport poet and reader beyond the here and now. It is this reaching beyond the particular that leads Darío to a cosmic vision in which the specific is subsumed and acknowledged as limited. His ideal love is simultaneously one and many, archetype of woman and representative of women. Her multifaceted nature is captured in the evocative couplings of "sea and cloud, foam and wave," which underscore both her spiritual and sexual dimensions. The erotic rhythms of the sea echo within Darío's poetry as language and poet unite, as desire is made flesh, as poetry approaches music.

Darío's journey ends with the evocation of a male voice that speaks with mystical overtones. When he suggests that his love sleep as he lights the censers, the quiet takes on a religious quality that is reinforced by the mention of a unicorn—associated with Christ—and dromedaries—summoning Biblical allusions. Darío thus broadens the nature of his undertaking. He strives to create a poetry that is simultaneously Spanish American and universal, that is, a poetry that melds and surpasses its artistic antecedents and provides a "spiritual" response to its context. As a result, the poet emerges as savior, a savior that finds transcendence in eroticism, spirituality in art, and profound knowledge in both.[19] This savior reappears in "Sonatina" ["Sonatina"], the next poem of the collection.

At the end of "Sonatina" the sad princess is given hope for happiness, love, life, and salvation with the impending arrival of a special beloved. Her fairy godmother tells her:

> "en caballo con alas, hacia acá se encamina,
> en el cinto la espada y en la mano el azor,
> el feliz caballero que te adora sin verte,
> y que llega de lejos, vencedor de la Muerte,
> a encenderte los labios con su beso de amor!" (*Poesía* 188)
>
> ["the joyous knight who adores you unseen
> is riding this way on his wingèd horse,
> a sword at his waist and a hawk on his wrist,
> and comes from far off, having conquered Death,
> to kindle your lips with a kiss of true love!"
>
> (Translation by Lysander Kemp 53)]

No matter how frivolous "Sonatina" appears at first, with its nursery-rhyme rhythm and its fanciful gardens and palace, by the final stanza the profound nature of the fairy-tale couple becomes evident.[20] The knight who arrives mounted on his winged steed, victor of Death, is more than the proverbial Prince Charming who appears in time to revive the lovesick princess. The linking of the hero-savior with Pegasus, the horse of the Muses, identifies the hero as an artist. His ability to lead his love, and his readers, out of the imperfect present into a paradisiacal future recalls the Christlike attributes that become a recurrent feature of Darío's later poetry about poetic responsibility.[21] If the knight that arrives at the end of "Sonatina" most closely corresponds to the poet-hero-savior, the princess that awaits him is the female consort of the male creator. She is the "flesh" of poetry, poetic language. It is here that the central image and the title of the poem show their fundamental union, one that is etymological. The rich, elaborately housed princess serves as a female other, a type of *muse*, who makes possible the creative eloquence of the male voice. She allows him to fulfill his role as savior by turning language into *music*.[22] This goal is further emphasized by the rhythmic virtuosity of the poem, which is written in superb dactylic alexandrines. The interplay among language, music, and poetry thus delineated in "Sonatina" is a continuation of Darío's attempts to clarify his artistic aims, which he had begun in "Palabras liminares," "Era un aire suave . . . ," and "Divagación." The poem begins with a revealing first stanza:

> La princesa está triste . . . ¿qué tendrá la princesa?
> Los suspiros se escapan de su boca de fresa,
> que ha perdido la risa, que ha perdido el color.
> La princesa está pálida en su silla de oro,
> está mudo el teclado de su clave sonoro;
> y en un vaso olvidada se desmaya una flor. (*Poesía* 187)
>
> [The Princess is sad. What ails the Princess?
> Nothing but sighs escape from her lips,
> which have lost their smile and their strawberry red.
> The Princess is pale in her golden chair,
> the keys of her harpsichord gather dust,
> and a flower, forgotten, droops in its vase.
>
> (Translation by Lysander Kemp 52)]

Darío holds that poetic language has lost its vitality and color; it is imprisoned in a golden vessel. The *music* that should be heard is silent; the atmosphere is stifling, unimaginative ("Parlanchina, la dueña dice cosas banales" ["the duenna prattles of commonplace things"]), and uninspired ("y, vestido de rojo, piruetea el bufón" ["the clown pirouettes in his crimson and gold"]). Poetry's only escape is through dreams of freedom and flight. "La princesa persigue por el cielo de Oriente / la libélula vaga de una vaga ilusión" ["the Princess traces the dragonfly course / of a vague illusion in the eastern sky"]. The grace and airiness of the dragonfly are contrasted with the princess's earthbound opulence. Though her possible lovers offer affluence, she longs for other things.

> ¡Ay! La pobre princesa de la boca de rosa
> quiere ser golondrina, quiere ser mariposa,
> tener alas ligeras, bajo el cielo volar,
> ir al sol por la escala luminosa de un rayo,
> saludar a los lirios con los versos de mayo,
> o perderse en el viento sobre el trueno del mar. (*Poesía* 188)

> [Alas, the poor Princess, whose mouth is a rose,
> would be a swallow or a butterfly;
> would skim on light wings, or mount to the sun
> on the luminous stair of a golden sunbeam;
> would greet the lilies with the verses of May,
> or be lost in the wind on the thundering sea.
>
> (Translation by Lysander Kemp 52)]

The princess aspires to an intense knowledge which will afford her direct access to the perfect and pure language of nature. She wishes to attain unmediated contact with the order of the cosmos. She rejects wealth and the reigning values of the day ("Ya no quiere el palacio . . ." ["She is tired of the palace . . ."]) because they interfere with her achieving the higher goal and greater pleasure of understanding the universe.

The objects that have come to be associated with the princess's imprisonment as well as with her physical and spiritual decline are boldly denounced. But Darío's detailed rejection is just the opposite of what it claims to be. It becomes a way of possessing, internalizing, and incorporating into his art those aspects that he pretends to disown.[23] He disdains the palace and its wealth as incapable of providing spiritual gratification. In fact they appear as obstacles to knowledge, as impediments that prevent individuals from seeing beyond the superficial trappings of life. At the same time, however, through his description he appropriates the luxuries that he repudiates.

This ambivalent position reflects a struggle that was common among *modernista* authors, who sought the praise and esteem of those whom they deemed superficial and insensitive. In "Sonatina," the poet challenges the materialism of bourgeois society. He strives to assert the worth of his creation in an environment that tends to ignore the value of his art, knowledge, and spiritual insight. The poet fights for the respect and esteem that he feels he deserves by taking up the weapons of the enemy—wealth and material comfort—and by poetically rendering them impotent. As the princess's riches are made subservient to the spiritual power offered by the poet, the value of poetic vision and artistic achievement is doubly raised above everyday reality—"the life and times to which [he] was born"—providing as clear a response to the established order as Martí's, Casal's, or Silva's. Only after the princess (poetic language) recognizes the appropriate (inferior) position of material riches can the poet fulfill his superior destiny.

"Sonatina" and the other initial pieces of *Prosas profanas* thus provide a bold, far-reaching statement of the *modernista* project. Darío, building on the positions taken by the early *modernistas,* forges a view of literature that assertively claims a crucial role within Spanish American society. *Modernismo* seeks to provide a vision of ultimate truths that would counter not only the positivist critique of religion but also the materialistic and pragmatic values predominant in modern life. It aspires to create a revitalized language appropriate to the new era that Spanish America had entered by embracing and transcending the beauty of art from across the ages and around the world. This language would resonate with the harmony of existence, reestablishing contact with the primal perfection of the universe and allowing the authors to reclaim the right to the moral compass of their countries. These goals, it is worth repeating, move from the philosophical, spiritual, and epistemological into the realm of values, priorities, and power. Perhaps no other poem more completely embodies the multifaceted thrust of *modernista* literature than "Coloquio de los centauros" ["Colloquy among the Centaurs"], the great masterpiece of *Prosas profanas.*[24]

The centaurs, composite creatures that embody the polar elements of intelligence and brutishness, are particularly poignant figures. Not unlike the poet, they aspire toward the divine but may fall toward the bestial. Their search both for the reconciliation of the tensions within themselves and for their reintegration into a harmonious and well-working universe provides the structuring mechanism through which Darío discusses the full range of possibilities embraced by the *modernista* movement. The poem begins on the magical Isla de Oro [Isle of Gold], where the centaurs discover the hidden order of the cosmos. Within this spiritual haven, they sense the harmony of existence, which provides the basis for the resolution of all strife. Through the comprehension of the orderly workings of the universe, the centaurs begin to perceive the proper role of sexual differences and mortality. They come to understand that the acceptance of the fundamental accord of all existence—the essential unity of the bestial and the divine, good and evil, male and female, life and death—is the key to the paradisiacal vision that they pursue. As a result, "Coloquio de los centauros" represents an idyllic interlude in the harried modern world. It presents a poetic depiction of a utopian alternative to the crass reality of contemporary life in which transcendent visions have been suppressed or ignored.

Quirón, the wisest of all centaurs, opens the colloquy with a discussion of the unity of all life and the unfading glory of the Muses. From the initial sections there arises a caldron of images in which harmony and art, creation and erotic passion, birth and death come together in a vision of supreme order in which the poet is guide and master.

> el vate, el sacerdote, suele oír el acento
> desconocido; a veces enuncia el vago viento
> un misterio; y revela una inicial la espuma
> o la flor; y se escuchan palabras de la bruma . . . (*Poesía* 201)

[the seer, the priest, usually hears the unknown / accent; at times the errant wind states / a mystery; and the foam or flower reveals / an initial; and words are heard in the foggy mist. . . .]

The poet knows what others fail to perceive. He can "read" the gesture, the sign, or the puzzle of external forms; he understands the language of nature and comprehends the order of the universe. It becomes his responsibility to translate the text of the world, revealing harmony where others only see—or create—discord.

Exactly how this obligation is best achieved is not fully delineated until 1901 and the publication of "Las ánforas de Epicuro" ["Epicurus's Amphorae"] (*Poesía* 234–241). This grouping of 13 poems, which was added along with "Cosas del Cid" ["About the Cid"] and "Dezires, layes y canciones" ["Sayings, Ballads, and Songs"] to the augmented edition of *Prosas profanas,* develops the neoplatonic and pantheistic worldview of "Coloquio de los centauros." These poems, among the best that Darío ever wrote, subtly and elegantly address Darío's most profound concerns. "La fuente" ["The Spring"] deals with the issue of poetic creation.

Joven, te ofrezco el dón de esta copa de plata
para que un día puedas clamar la sed ardiente,
la sed que con su fuego más que la muerte mata.
Mas debes abrevarte tan sólo en una fuente.
Otra agua que la suya tendrá que serte ingrata;
busca su oculto origen en la gruta viviente
donde la interna música de su cristal desata,
junto al árbol que llora y la roca que siente.
Guíete el misterioso eco de su murmullo;
asciende por los riscos ásperos del orgullo,
baja por la constancia y desciende al abismo
cuya entrada sombría guardan siete panteras;
son los Siete Pecados, las siete bestias fieras.
Llena la copa y bebe: la fuente está en ti mismo. (*Poesía* 235)

[Young man, I offer you the gift of this silver goblet / so that one day you may calm your burning thirst, / the thirst whose fire kills more than death. / But you must drink only from one fountain.

Water other than from it will fail to satisfy you; / look for its hidden source in the living grotto / where the internal music of its crystal breaks free, / next to the tree that cries and the rock that feels.

Let the mysterious echo of its murmur guide you; / climb up through the harsh rocks of pride, / go down through constancy and descend to the abyss

whose dark entrance seven panthers guard; / they are the Seven Sins, the seven wild beasts. / Fill the goblet and drink: the spring is in you yourself.]

Poetry is the silver goblet that the young man will use to calm his burning passion to create and, through artistic creation, to provide a transcendent vision for those around him. But his longing will be satisfied only when the goblet is filled with the water that flows within himself. This internal spring is the part of him that resonates with the primordial language of music and the primal order of existence. Its entrance may be blocked by habit, custom, or any number of corrupting forces, but he must strive to reach it. Though inevitably the music of the spring will be muted in the goblet, the poet must fill poetic form with that purest, most responsive, and most sincere part of himself.

Despite the powerful presence of these declarations of poetic intent,

Prosas profanas has often been characterized by its brilliant images, musicality, and aesthetic play. It has been judged in terms of its formal artistry and its discovery or recovery of a large variety of verse forms begun by earlier *modernistas.* In addition to the dactylic alexandrine of "Sonatina," Darío resuscitated verses of twelve syllables in "Elogio de la seguidilla" ["Praise of the Seguidilla"] and the poetry of the *cancioneros* of the fifteenth century in "Dezires, layes y canciones." Through caesuras placed at different points and the use of enjambment, he further expanded the musical effect of traditional forms. The great flexibility of poetic prose appears in "El país del sol" ["The Land of the Sun"], and the synesthetic mixing of music and color is evident in "Sinfonía en gris mayor" ["Symphony in Gray Major"], a poem inspired by the Parnassian art of Gautier.[25] Yet behind the artistic experimentation and open eroticism, one finds a serious search for linguistic, artistic, conceptual, and spiritual freedom. Darío would continue this search as he took up residence on the other side of the Atlantic.

After living in Buenos Aires for five years, Darío returned to Spain in 1898 as correspondent for *La Nación.* He was to remain in Europe for several years, residing in Barcelona, Madrid, and Paris. He also traveled in Italy and southern Spain, finding much to admire throughout his journeys. Between 1902 and 1905 he wrote a number of articles, which he later published as *Opiniones* [*Opinions*] (1906). In 1906, after a trip to Brazil, he was named Nicaraguan consul in Paris and then, in 1907, Nicaraguan ambassador to Spain. Though during these years he was at the height of his career and fame, he never fully resolved the economic and personal problems that plagued him from the beginning. He did, however, find a supportive and caring female companion, Francisca Sánchez, who, in 1900, bore him a daughter, who died immediately, and, in 1903, a son, whom he nicknamed Phocás and referred to in a poem published in his next major work, *Cantos de vida y esperanza.* Tragically, Phocás lived just two years. A second son, also called Rubén Darío Sánchez, was born in 1907.

In *Cantos de vida y esperanza. Los cisnes y otros poemas* [*Songs of Life and Hope. The Swans and Other Poems*] (1905), Darío reveals himself to be a poet more sure of himself and more willing to express his sense of difference—his sense of being Spanish American. The imported models that dominated his poetic imagination in *Prosas profanas* recede to the background, and Darío begins to speak in a more somber, mature voice. The poems reflect on the passage of time and on the youthful squandering of energies. They also comment directly on the sociopolitical context that up to this point remained, for the most part, veiled in metaphor.

Like *Prosas profanas, Cantos de vida y esperanza* begins with an important prose introduction that responds to the critics of the day. Darío defends his poetry by emphasizing its grounding in the "aristocracy of thought" and the "nobility of art" which, in turn, continue to be offered as antidotes to the mediocrity, intellectual stultification, and aesthetic superficiality that he sees in contemporary society. With pride he recognizes that *modernismo,* unlike previous literary movements, originated in Spanish America and subsequently spread to Spain. While his claim to be the founder, rather than the head, of the movement is misleading, Darío certainly gave to *modernismo* a breadth of vision, a philosophy of language, and an intellectual depth that provided writers with an overarching and unifying framework. He also increased, as he notes, despite entrenched resistance, the flexibility of Spanish poetic expression, maintaining as his goal a pure and direct reflection of his view of beauty.

In this introductory statement, Darío reiterates his ties to romantic literary theory and its insistence that the artist must remain true to himself. As early as 1882, Martí had expressed a similar commitment to sincerity and integrity, one that was not simply personal or aesthetic in nature. For Martí, being faithful to himself entailed presenting his positions related to cultural identity and national autonomy openly and imaginatively. While this aspect of the *modernista* project had usually been addressed indirectly in Darío's writings, it takes on new urgency with *Cantos de vida y esperanza,* a legacy of the Spanish American War of 1898 and U.S. intervention in the "creation" of Panama in 1903. He concludes the preface with a powerful statement:

> If in these songs there is politics, it is because it is universal. And if you find verses to a president, it is because they echo a continental clamor. Tomorrow we could all be Yankees (and it is the most likely); at any rate, my protest remains inscribed on the wings of immaculate swans, as illustrious as Jupiter. (*Poesía* 244)[26]

At this moment in his career, Darío sets forth with renewed impetus to reaffirm Victor Hugo's ideal of poetic responsibility (first alluded to in *Azul . . .*). For Hugo, the poet's mission is compulsory, to be performed with divine sanction and divine inspiration. It encompasses a commitment that is both spiritual and political, a moral obligation to understand the world and to improve it (Shroder 67–68). It is therefore significant that Darío chooses to reproduce Hugo's "Celui qui . . ." in the first words of the opening poem ("Yo soy aquel que ayer no más decía . . .") ["I am the one who only yesterday was saying . . ."] as he reassesses his life up to this point.

Darío's turn to politics is best observed in "Salutación del optimista" ["The Optimist's Greetings"] (*Poesía* 247–248), the second poem of the volume and a "political" complement to the more introspective and "spiritual" initial piece, as well as in "A Roosevelt" ["To Roosevelt"] (*Poesía* 255–256), the eighth poem of the collection. Both compositions express concern about the power of the United States and the vulnerability of Spanish America, "Salutación del optimista" obliquely and "A Roosevelt" directly. "A Roosevelt" presents Spanish America as having one asset that offsets all the rest: its spiritual strength. While this inner strength is summarized by the final line of verse defiantly addressed to Roosevelt, "Y, pues contáis con todo, falta una cosa: ¡Dios!" ["And, then you count on everything, one thing is missing: God!"], it is more fully explored in the section that begins: "Mas la América nuestra . . ." ["But our America . . ."]. As Keith Ellis has observed, "The first fifteen verses of this second part of the poem establish the traditional idealism of the Spanish American people. The first of the attributes that the poet finds in Spanish America is that, in contrast to the United States, 'tenía poetas desde los viejos tiempos de Netzahualcoyotl' ['it had poets since the ancient times of Netzahualcoyotl']" (98). The wisdom of its poets and the spiritual fortitude of its ancestors provide Spanish America with the means to counter the "explosive" materialistic forces of modernity: "energy," "progress," strength ("el culto de Hércules" ["the cult of Hercules"]), and greed ("el culto de Mammón" ["the cult of Mammon"]). The poet's "optimism," evident in "Salutación del optimista," derives from his faith in his people to maintain—under his tutelage—past traditions and to renew contact with a transcendental realm. This knowledge has the capacity to empower and transform and therefore entails political praxis.

Despite assertions of optimism in "Salutación del optimista" and in the title *Cantos de vida y esperanza*, the collection in general wavers between hope and despair. Darío confronts the passage of time and the inevitability of death ("Canción de otoño en primavera" ["Autumn Song in Spring"], "A Phocás el campesino" ["To Phocás the Farmer"]), and he struggles with religious doubts and despair. When traditional beliefs fail him, he finds consolation in his faith in art, in the harmony of the universe, and in the perfectibility of man ("Mientras tenéis, oh negros corazones . . ." ["While you have, oh black hearts . . ."], "Helios" ["Helios"], "Filosofía" ["Philosophy"], "Ay, triste del que un día . . ." ["Pity for him who one day . . ."], "Caracol" ["Seashell"]). Even love becomes an aspect of his search for transcendence. Influenced by esoteric thought, Darío's erotic poetry is no longer as playful, light-hearted, or defiantly rebellious as it had been throughout most of *Prosas profanas;* it now evokes the eternal order and perfection of creation ("Por un momento, oh Cisne, juntaré mis anhelos . . ." ["For one moment, O Swan, I will unite my desires" . . .], "¡Antes de todo, gloria a ti, Leda! . . ." ["Before all else, glory to you, Leda! . . ."], "¡Carne, celeste carne de la mujer! . . ." ["Flesh, celestial flesh of woman! . . ."], "En el país de las Alegorías . . ." ["In the land of Allegories . . ."], "Amo, amas" ["I Love, You Love"], "Programa matinal" ["Morning Program"]). Yet the anarchy of modern life repeatedly cuts him off from this sense of well-being and belonging, leaving him either to recall the religious answers of his youth ("Canto de esperanza" ["Song of Hope"], "Spes" ["Spes"], "¿Qué signo haces, oh Cisne, con tu encorvado cuello . . ." ["What sign do you make, O Swan, with your curved neck . . ."], "¡Oh, terremoto mental! . . ." ["O mental earthquake! . . ."], "El verso sutil que pasa o se posa . . ." ["The subtle verse that passes or that rests . . ."]) or to suffer anguish, despair, and guilt ("La dulzura del ángelus" ["The Sweetness of the Angelus"], "Nocturno I" ["Nocturne I"], "Melancolía" ["Melancholy"], "Nocturno II" ["Nocturne II"], "Lo fatal" ["Fatality"]).

Unable to see beyond the chaos and disorder around him, he accuses himself of failing to fulfill the divine destiny he claimed for himself as poet-seer in "Alma mía" ["My Soul"] of "Las ánforas de Epicuro." Having set for himself the quasi-religious goal first formulated among the romantics to "redeem man by fostering a reconciliation with nature" (Abrams, *Natural Supernaturalism* 145), Darío's sense of failure generates a reaction not unfamiliar to contemporary readers. From the early romantics to the present, writers have confronted the fragmentation, estrangement, and alienation of modern man. The emotional intensity of Darío's response retains a fresh quality that affirms a spiritual continuity between *modernismo* and contemporary perspectives. This struggle between hope and despair comes to the fore in *Cantos de vida y esperanza* in the two "Nocturnos" and "Lo fatal."

In the first "Nocturno" (*Poesía* 270), Darío announces the central theme of these three poems: the dual horror of consciousness and conscience. He confronts the fleeting nature of existence, the halting but inexorable march toward "the unavoidable unknown," and the disjunction between the artistic and personal goals he has set for himself and what he has actually achieved. The distant clavichord never yielded its sublime sonata to the poet's imagination, and he now fears that he must pay the cost of his search for beauty and pleasure. His only consolation is the belief that life is merely a nightmarish, fitful sleep from which he will be awakened to see a truer reality. This image is taken up in the second "Nocturno."

> Los que auscultasteis el corazón de la noche,
> los que por el insomnio tenaz habéis oído
> el cerrar de una puerta, el resonar de un coche
> lejano, un eco vago, un ligero ruido . . .
> En los instantes del silencio misterioso,
> cuando surgen de su prisión los olvidados,
> en la hora de los muertos, en la hora del reposo,
> ¡sabréis leer estos versos de amargor impregnados! . . .
> Como en un vaso vierto en ellos mis dolores
> de lejanos recuerdos y desgracias funestas,
> y las tristes nostalgias de mi alma, ebria de flores,
> y el duelo de mi corazón, triste de fiestas.
> Y el pesar de no ser lo que yo hubiera sido,
> la pérdida del reino que estaba para mí,
> el pensar que un instante pude no haber nacido,
> ¡y el sueño que es mi vida desde que yo nací!
> Todo esto viene en medio del silencio profundo
> en que la noche envuelve la terrena ilusión,
> y siento como un eco del corazón del mundo
> que penetra y conmueve mi propio corazón. (*Poesía* 291)

> [You that have heard the heartbeat of the night,
> you that have heard, in the long, sleepless hours,
> a closing door, the rumble of distant wheels,
> a vague echo, a wandering sound from somewhere:
> you, in moments of mysterious silence,
> when the forgotten ones issue from their prison—
> in the hour of the dead, in the hour of repose—
> will know how to read the bitterness in my verses.
> I fill them, as one would fill a glass, with all

> my grief for remote memories and black misfortunes,
> the nostalgia of my flower-intoxicated soul
> and the pain of a heart grown sorrowful with fêtes;
> with the burden of not being what I might have been,
> the loss of the kingdom that was awaiting me,
> the thought of the instant when I might not have been born
> and the dream my life has been ever since I was!
> All this has come in the midst of that boundless silence
> in which the night develops earthly illusions,
> and I feel as if an echo of the world's heart
> had penetrated and disturbed my own.
>
> (Translation by Lysander Kemp 87–88)]

If life is a fitful sleep, the nights of insomnia become the moments of vision. It is during the dark, sleepless hours that Darío sees with greatest clarity both the illusion of life and the omnipresence of death. He joins with all who, in their sleepless self-reflection, have developed an acute sensitivity to the world that surrounds them. They are the ones who, in the mysterious silence of the night, when the past escapes from the prison of oblivion and resurfaces as the voice of conscience, understand the full significance of his verse. By the fourth and penultimate quatrain, Darío reveals that his greatest concern is that he may have failed to be what he should have been and that he has lost the kingdom that should have been his. While the poet's harmony with the beat of universal life implies the promise of salvation, it also reminds him of the responsibilities of his vocation and, as is even more evident in "Lo fatal," reinforces the imperatives of the doctrine of transmigration of souls. If his is an elevated soul, as his poetic sensibility would suggest, he must fulfill the obligations of that exalted status. If he does not, his soul will descend on the scale of existence during future incarnations. The anguish that results from this possibility is most intensely expressed in the fourth stanza. The alliteration of the *p*'s emphasizes the echo of "pesar" ["sorrow" or "regret" but also the verb "to weigh," "to be weighty"] in "pensar" ["to think"] and evokes the overwhelming impact of the poet's loss ("pérdida"), while the masculine rhyme of the second and fourth lines intensifies the poet's cry of remorse for his mistakes. With its deliberately ambiguous ending, this second "Nocturno" captures Darío's seesawing emotions, his sense of inadequacy and despondency as well as his hopes and pride.

In "Lo fatal," the last poem of *Cantos de vida y esperanza*, Darío juxtaposes the sense of failure to meet his spiritual responsibilities with the apparent ignorance or insensitivity of lower forms of life. While he envies the insusceptibility of animals and plants to the pangs of conscience, which appear here as products of heightened consciousness, Darío dreads the thought that his soul may descend to a subhuman level in retribution for having sullied his elevated status with "the flesh that tempts." This allusion to the great chain of being enhances the poem's evocation of downward spiraling despair. By the end, there is a total breakdown of grammatical and strophic structures. The poet conveys the stranglehold of fear with sentences and stanzas that remain unfinished, with the purposeful elimination of the last line of what would have been a sonnet. The poem thus becomes emblematic of the course of the *modernista* project as well, for as alienation and existential dread block access to the desired transcendental vision of universal perfection, *modernismo* moves away from formal beauty to a more experimental conception of art.[27]

Spanish American literature does not, however, turn away from the aggressively antiempiricist stand taken by the *modernistas*, which is epitomized by their recourse to occultism. Throughout the poems just discussed, the alternative way of knowing embodied in esoteric tradition lurks in the background. These beliefs, as I have shown in my *Rubén Darío and the Romantic Search for Unity*, move to the forefront in many of Darío's works and become essential elements in the way he came to envision the world. The incorporation of occultist tenets within Darío's poetic universe reflects his dissatisfaction with the limited conceptions of reality accepted by mainstream, bourgeois society. This rejection of the prevailing positivistic perspective endures well into the twentieth century. So does the recourse to esoteric beliefs and images. Among the later *modernistas*, most notably Nervo, Lugones, and Herrera y Reissig, occultism continues to offer an enhanced perspective on the world and opens the possibility of achieving harmony, unity, and accord. For those of the avant-garde and after, the focus is on the abiding renunciation of established beliefs, the taken-for-granted, and all that appears "routine," "natural," and "obvious." Among the postmoderns, the challenge ultimately extends to foundational philosophies. Darío's concerns about the power of predominant epistemological perspectives reappear in the prose introduction to his next collection of verse, *El canto errante* [*The Wandering Song*] (1907).

The prose introduction, "Dilucidaciones" ["Elucidations"] (*Poesía* 299–306), was originally written as a response to a request by *El Imparcial* for a clarification of the fundamental tenets of *modernista* art. The backdrop to Darío's reply, like Martí's prologue to *El poema del Niágara* (discussed in chapter 2) is the modern world filled with cars and bombs (*Poesía* 299) and holding little sympathy for poets (300). Like Martí, Darío rejects the "precepts, pigeonholes, customs, and clichés" that would limit the freedom of those who would leave the past and enter the future (301). It is precisely the future that Darío anticipates. He recognizes that language determines thought and that concepts limit vision. He writes: "The verbal cliché is harmful because it encloses within itself the mental cliché, and together they perpetuate paralysis, immobility" (302).[28] In this simple but powerful statement, Darío proposes a role for art that goes beyond *modernista* tenets. Art, free from the structural constraints of social or scientific discourse, offers a corrective, one that can liberate vision and open possibilities otherwise unperceived. This confidence in art carries through into the avant-garde and leaves its impact upon postmodern sensibilities even after the freedom of the artist is put in question. For Darío, however, his rejection of linguistic and conceptual "molds" is tied to a more traditional, romantic attitude that his art is born of a natural accord between "body and soul" (language and inspiration) that reflects an even more profound harmony achieved through contact with "the vast universal soul" (*Poesía* 304). Darío concludes the introduction by attributing to the poet the ability to attain this totalizing vision: "The true artist understands all modes and manners and finds beauty in all forms. All glory and all eternity are in our conscious grasp" (*Poesía* 306).[29] This final assertion, perhaps inadvertently, also points to the volume's most salient feature, namely, its diversity of themes, styles, and verse forms.[30]

In 1907 Darío went home to Nicaragua, where he was accorded all the honors of a national hero. He was named Nicaraguan ambassador to Spain and thereby secured a steady if modest income. Upon his return to Madrid, he was received again with adulation and honors. From then until 1914 Darío spent most of his time in Spain and France, though he did take trips to Mexico, Brazil, Uruguay, and Argentina. In 1910 he was asked to write a poem commemorating the hundredth anniversary of Argentina's independence. The resulting "Canto a la Argentina" ["Song to Argentina"] was published in *La Nación* on May 25 and later became the centerpiece of *Canto a la Argentina y otros poemas* (1914). The commissioned work turned out to be Darío's longest single poem, a masterpiece of civic poetry that reveals hidden links to other Spanish American literature. Its vision is wide-reaching and all-inclusive, moving freely from scenes from Greek mythology to the wheat fields of the Pampas to the latest immigrants seeking solace and sustenance in their new homeland. While some sections are patriotic and grandiloquent, others are intimate and lyrical. Its overall exuberance, however, is conveyed by an abundance of images and detailed, elaborate description. All these elements, the tone, the images, and, to a certain extent, the themes recall Oviedo's *Historia general y natural de las Indias,* Zequeira y Arango's "A la piña," Bello's "La agricultura de la zona tórrida," and Lugones's *Odas seculares.* It even anticipates sections of Neruda's *Canto general* with its panoramic perspective, its inclusion of the common man, and its search for a new Spanish American identity. "Canto a la Argentina," however, is unerringly optimistic, a national reflection of Darío's personal aspiration to harmony and accord through beauty and high culture. Its enthusiastic endorsement of Argentina reflects once again the *modernista* ambivalence to the society that both marginalized and idealized its poets. Whereas Darío's politics are compromised by need and desire, those that follow in his footsteps—Neruda, Vallejo, and others—turn his visionary perspective into a confrontational reexamination of mores, values, and history.

During the last years of his life, Darío published two volumes of verse, *Poema del otoño y otros poemas* [*Autumn Poem and Other Poems*] (1910) and *Canto a la Argentina y otros poemas* [*Song to Argentina and Other Poems*] (1914), as well as four volumes of prose. *Letras* [*Letters*] (1911) and *Todo al vuelo* [*All on the Run*] (1912) bring together articles written from 1906 to 1909. Darío also wrote two revealing biographical works: *La vida de Rubén Darío escrita por él mismo* [*The Life of Rubén Darío Written by Darío Himself*] (1912) and *Historia de mis libros* [*History of My Books*] (1913). In 1914 World War I broke out and Darío had to leave Paris. Ailing and without means, he found shelter in Guatemala thanks to the hospitality of that country's dictator, Manuel Estrada Cabrera. In 1915 Rosario, his legal wife, came for him and took him back to Nicaragua, where, in 1916, after two operations, he succumbed to years of self-inflicted abuse, but not until altering forever the trajectory of poetry written in Spanish.

FIVE

Continuity within an Evolving Movement

It is impossible to determine a specific moment at which *modernismo* reached its plenitude. Somewhere between 1896 and 1905—between the publication of *Prosas profanas* and *Cantos de vida y esperanza,* that is, during the period in which Darío moved from Buenos Aires to Madrid—*modernismo* developed a strong sense of identity as a movement and reached its widest diffusion. The shift that appeared in Darío's poetry during this time was symptomatic of the changes showing up in the most creative and effective poetry throughout the continent. Though this "second stage" of *modernismo* has been viewed as encompassing the heterogeneous group of *modernista* poets who survived beyond 1896, the close of the movement has been in question for years. Focusing on changes in tone, style, and images, literary historians sought to distinguish between works that were continuations of *modernismo* and those that manifested departures from the movement. Some authors were called late *modernistas,* while others were labeled *posmodernistas.* Since early critics had defined *modernismo* as fundamentally preoccupied with aesthetics, they failed to realize that the formal variations that they observed did not constitute an essential alteration of the *modernista* stance. For the most part, the late *modernistas* as well as the so-called *posmodernistas* persisted in efforts to comprehend and express the nature of existence and merely continued the simplifying trends begun within *modernismo* itself. Moreover, like the earlier *modernistas,* they remained confident that the knowledge that they could provide would influence the spiritual, social, and political life of their countries, all of which were confronting the diverse consequences of modernization.

To a certain extent the classification *posmodernismo* sidesteps an assessment of what constitutes central rather than peripheral features of this literature. As "a middle generation," this group of writers has been defined by its distancing from *modernismo* but, at the same time, by its refusal to break with traditional belief systems and to cross the threshold into the more defiant avant-garde. Forster's introduction to this "ten-year generation" reinforces its transitional nature.

> . . . all the poets from this intermediate moment accepted *modernismo* as a point of departure, in order to modify it or reject it later, each one of them following his own criterion. Some never break completely with the previous thematics and form; others rapidly go toward the experiments of the avant-garde or toward the American realities of *criollismo.* (99) [1]

The decisive shift that actually signals an essential reorientation away from the *modernista* agenda comes when the overall optimism of *modernismo* gives way to a more pervasive doubt and the abiding faith in a divine order is eroded by penetrating anguish. Confidence that cultural patterns are capable of revealing, evoking, or reflecting transcendental truths is undermined by a corrosive skepticism. The pursuit of formal beauty and logic recedes as art focuses on and aggressively confronts what is interpreted to be an increasingly hostile environment. This profound alteration is what defines *la vanguardia,* the avant-garde.

In this chapter, I focus on those late *modernista* poets that continued and advanced the predominant *modernista* tendencies, some of whom have on occasion been classified as *posmodernistas.* They include Enrique González Martínez, Amado Nervo, Ricardo Jaimes Freyre, Guillermo Valencia, José María Eguren, and José Santos Chocano. In the next chapter, I will discuss those late *modernista* poets that drew upon the *modernista* impetus toward change and pushed further the expansion of the poetic repertoire in Spanish, thereby anticipating the movement's transformation into the Hispanic avant-garde. This tendency emerges in the works of Leopoldo Lugones, Julio Herrera y Reissig, and Delmira Agustini.

If Herrera y Reissig and Lugones are now recognized as poets who planted the seeds that bore fruit during the creative explosion of *la vanguardia,* González Martínez has long been linked with the poetic changes identified with *posmodernismo*—in the strictly Spanish American sense of the word. Yet the identification of González Martínez with a separate movement is misleading. From the perspective put forth in the previous chapters, the work of no other poet better shows that *posmodernismo* is not an independent movement but rather a continuation and natural outgrowth of *modernismo.* Throughout his long life and nineteen collections of verse, his artistic production reveals a continuity rather than a break with the predominant *modernista* characteristics. In poem after poem, González Martínez explores the fundamental *modernist* desire to reveal the hidden order of the universe through the grace, beauty, and harmony of poetry.

Early critics who saw in his work a departure from *modernismo* tended to view *modernista* verse as superficial, decorative, concerned only with the formal enhancement of Spanish poetry. Many failed to recognize (1) that *modernismo*'s recourse to swans, gardens, princesses, palaces, and European cultural models of all sorts reflected serious concerns regarding language, reality, and Spanish America's place in the modern world, and (2) that well before 1911, as early as the publication of "Las ánforas de Epicuro" in the 1901 edition of *Prosas profanas, modernismo* had moved toward a simpler, more introspective, and self-assured style. Early critics also misunderstood the intended meaning of "Tuércele el cuello al cisne . . ." ["Twist the neck of the swan . . ."]. Contrary to their contention, the "swans of deceitful plumage" to which González Martínez refers did not belong to *modernista* poets but rather, as he himself would make clear, to the myriad, now long-forgotten, hack imitators who echoed the language of *modernismo,* its opulence, elegance, and ornamentation, without comprehending the underlying issues that defined *modernista* poetics. In his autobiographical text, *La apacible locura* [*Peaceful Insanity*], González Martínez contrasts the differences between *modernista* verse and distortive imitations and asserts his admiration for the fundamental spirit of the movement.

In reality the poem only . . . expressed a reaction against certain *modernista* commonplaces drawn from Rubén Darío's opulent lyrical baggage, the Darío of *Prosas profanas* and not the Darío of *Cantos de vida y esperanza.* Leaving aside what is essential in the poetry of the great Nicaraguan, what continued in his imitators was what we could call appearance and procedure. Of course the grace, the exceptional virtuosity,

> and the charming personality of their model were missing among the imitators. Nor did Darío's followers achieve his lyrical emotion, perceptible in him from *Prosas profanas* on, even in poems where the technical agility and the control of form seemed to be the only creative intention; much less the emotion that Rubén's poetry would achieve in *Cantos de vida y esperanza*, already complete, mature, and wise. The only thing that was accessible to the imitators was the themes—swans, pages, princesses; the metrics—already taken from French or from old Spanish poetry; the use of adjectives, that by dint of repetition lost efficacy and novelty; in general, the word, sterile to those that steal it, and not the spirit, fertile and rejuvenating. (159)[2]

As can be inferred from this statement, González Martínez preferred a simpler and less adorned language to capture the sympathetic resonance between the soul of the poet and the soul of the world. His faith in the power of poetry to achieve this end was grounded in the pantheism, occultism, and aesthetics of the earlier *modernistas* and symbolists, many of whom he had translated.[3]

González Martínez's poetry stands apart from the irony, skepticism, and doubt of avant-garde poets. It is firmly anchored instead in the *modernista* foundation of analogy, revealing the movement's true nature. *Modernismo*'s profound significance was and remains hidden from readers who fail to realize that *modernista* writings seek to penetrate—through the musicality and evocative power of its art—the eternal and harmonious order of existence that, in turn, is concealed by the chaos of everyday reality. *Modernismo*'s lasting legacy is precisely the aspiration to this "hidden meaning" and the hope that, through its revelation, art can have consequential repercussions.

The desire to set things straight and realign poetry with its far-reaching goals provides the background to González Martínez's most famous poem, "Tuércele el cuello al cisne . . . ," which was originally published in *Los senderos ocultos* [*The Hidden Paths*] in 1911. A series of commands to future writers structures the poem.

> Tuércele el cuello al cisne de engañoso plumaje
> que da su nota blanca al azul de la fuente;
> él pasea su gracia no más, pero no siente
> el alma de las cosas ni la voz del paisaje
> Huye de toda forma y de todo lenguaje
> que no vayan acordes con el ritmo latente

de la vida profunda . . . y adora intensamente
la vida, y que la vida comprenda tu homenaje.
Mira al sapiente buho cómo tiende las alas
desde el Olimpo, deja el regazo de Palas
y posa en aquel árbol el vuelo taciturno . . .
El no tiene la gracia del cisne, mas su inquieta
pupila, que se clava en la sombra, interpreta
el misterioso libro del silencio nocturno. (*Obras completas* 116)

[Wring the neck of the swan of deceitful plumage / that gives its white note to the blue of the spring; / it seeks only to show off its grace, but does not feel / the soul of things nor the voice of the landscape.

Flee from all forms and all language / that do not proceed in accord with the latent rhythm / of the profound life . . . and adore intensely / life, and may life understand your homage.

Look at the wise owl how it extends its wings / from Olympus, it leaves the lap of Palas / and rests its taciturn flight in that tree. . . .

It does not have the grace of the swan, but its restless / eye, that penetrates the darkness, interprets / the mysterious book of the nocturnal silence.]

In this beautifully executed sonnet, the *posmodernista* owl replaces the *modernista* swan. Whereas its natural grace embodied for Darío and other *modernistas* the rhythm of the universe and the perfection of form, the swan came to be identified with hollow echoes of *modernista* poetics and, therefore, with obstacles encumbering the vision of harmony—with deceit, obfuscation, and the inability to "hear the soul of things and the voice of the landscape." The owl does not have the swan's splendor, but it does have the ability to see into the dark and to interpret what others cannot detect. Formal beauty is thus subordinated to clarity of perception and an intuitive understanding of the order of the cosmos. For González Martínez, it is this combined ability to perceive and understand that shapes his poetry. Rather than a showy, superficial attractiveness, he seeks a profound grace that derives from the natural order of things.

Like the Darío of "Las ánforas de Epicuro" and *Cantos de vida y esperanza,* González Martínez hopes that his poetry will reflect the rhythms of existence, the art of nature, and the harmonious soul of the universe. But while Darío's poetry is energized by a tortured uncertainty, González Martínez's is characterized by a relentless optimism and a supreme confidence in the order of things. Darío confronts the gulf that exists between what is desired and what is real; he struggles to replace doubt and anguish with faith and solace; he suffers with the fear that he is not up to the challenge of his mission as he finds himself tempted by the society that he often condemns. These tensions are virtually absent from González Martínez's work and, accordingly, his individual poems become variations on a theme, masterful reworkings of one fundamental premise. "Irás sobre la vida de las cosas . . ." ["You will move over the life of all things . . ."] from *Silénter* [*Silently*], published in 1909, concludes with the following statement of poetic objectives:

> Y que llegues, por fin, a la escondida
> playa con tu minúsculo universo,
> y que logres oír tu propio verso
> en que palpita el alma de la vida. (*Obras completas* 70)

[And may you arrive, at last, at the hidden beach / with your minute universe, / and may you succeed in hearing your own verse / in which beats the soul of life.]

Similarly, he ends "Busca en todas las cosas . . ." ["Look in all things for . . ."] from *Los senderos ocultos* with a clear imperative:

> Busca en todas las cosas el oculto sentido;
> lo hallarás cuando logres comprender su lenguaje;
> cuando sientas el alma colosal del paisaje
> y los ayes lanzados por el árbol herido . . . (*Obras completas* 99)

[Look for the hidden meaning in all things; / you will find it when you succeed in understanding its language; / when you sense the colossal soul of the countryside / and the sighs hurled by the wounded tree. . . .]

In "Viento sagrado" ["Sacred Wind"], the first poem of *El libro de la fuerza, de la bondad y del ensueño* [*The Book of Strength, Goodness, and Fantasy*] (1917), González Martínez once again presents his faithful vision, this time in the face of the horrors of the Mexican Revolution. The "sacred wind" of the title refers to the powers of harmony and reconciliation that exist within a world temporarily gone awry.

Hará que los humanos,
en solemne perdón, unan las manos
y el hermano conozca a sus hermanos.
No cejará en su vuelo
hasta lograr unir, en un consuelo
inefable, la tierra con el cielo;
hasta que el hombre, en celestial arrobo,
hable a las aves y convenza al lobo; . . . (*Obras completas* 148)

[It [the sacred wind] will make humans, / in solemn forgiveness, unite their hands / and brother recognize his brothers.
It will not restrain its flight / until it succeeds in uniting, in ineffable consolation, / heaven and earth;
until man, in celestial rapture, / speaks to the birds and convinces the wolf; . . .]

The exceptions to this replay of themes and symbols (the night, the lake, the wind, the hidden fountain, etc.) appear in his final works, *El diluvio del fuego* [*The Deluge of Fire*] (1938) and *Babel* (1949). Poems in both collections address the death of his wife and son—the poet Enrique González Rojo—as well as gripping historical events. But even in these later poems that deal with personal loss and the horrors of Nazi Germany, González Martínez, in characteristic fashion, counters despair with trust and faith.

Amado Nervo, on the other hand, is both more like Darío and more representative of modern sensibilities, which assert an uneasy—even tortured—relationship with dominant beliefs and values. He has been characterized as a poet pulled between opposite poles—between a desire for material pleasures and an aspiration toward spiritual goals, between sensuality and religiosity, and between faith and doubt. His overwhelming longing to see transcendence in what appears limited and mutable, to reach beyond the immediate and the tangible, and to find a philosophic framework that would be consonant with his erotic nature led him, like Darío and the French symbolists before him, to explore a wide variety of unorthodox belief systems. He turned to ideas drawn from pantheism, mysticism, theosophy, spiritualism, Bergsonian vitalism, Buddhism, and Hinduism, and he found in them an effective alternative to the dry intellectualization of Spanish American positivism, which attached value exclusively to material things and endowed industry with glories and virtues. Nervo came to see the greatest failure of positivistic thought in its inability to grasp the nature of anything other than the purely mechanical and static.[4] Consequently, he sought to immerse himself in the flow of existence and to thereby achieve a knowledge of reality more profound than the fragmented and incomplete visions afforded by modern science and commerce. His *La hermana agua* [*Sister Water*] (1901?) can be read as a reaffirmation of the primacy of the "fluidity" of life (2:1380–1386). Like Martí and Darío, he asserts that to dictate form would be to inhibit what the poet most hopes to achieve, namely, direct contact and accord between language and the universe.

At the same time that Nervo explored unorthodox worldviews, he remained tied to his early religious training. His short-lived studies for the priesthood combined with his philosophic curiosity to reinforce the syncretic tendencies prevalent among *modernista* writers. He easily equates Christ with other divinities—Jove, Allah, Brahma, Adonai—and aspires to a loss of self, of self-importance, and of desire that is reminiscent of both Christian asceticism and Buddhist spiritualism. "Renunciación" ["Renunciation"] from *Serenidad* [*Serenity*] (1914) epitomizes this position: ". . . el edén / consiste en no anhelar, en la renunciación / completa, irrevocable, de toda posesión" [". . . Eden / consists of not desiring, of the complete, / irrevocable renunciation of all possessions"] (2:1602).

It is the recognition of these fundamental tensions in Nervo's work that has generated renewed interest in and a recent reassessment of his work. For a period of some forty years following his death, Nervo's reputation declined and critics questioned the value of his poetic production, faulting his writing for its supposed vulgarity, superficiality, and lack of originality. It may be true that some poems of his later collections suffer from a facile application of various philosophic perspectives or from an artificial cultivation of a sense of intimacy. Yet his poetic output demonstrates tremendous range. It moved from a mastery of *modernista* aesthetics, with its aspiration to grace, elegance, and richness of texture, to a controlled, personal, and intimate style. All the while, he addressed diverse issues of fundamental concern—from the political and social to the personal and philosophic—returning repeatedly to the universal themes of time, change, loss, love, and desire, all of which come together in his much quoted "Vieja llave" ["Old Key"] from *En voz baja* [*In a Soft Voice*] (1909) (2:1558–1559).

"Vieja llave" undertakes a sensitive examination of the power of evocation in a world of flux and impermanence. The poem's greatest concern reappears in variations on the following question: "si no cierra ni abre nada, / ¿para qué la he de guardar?" ["if it does not open or close anything, / why should I hold on to it?"]. The chiseled key ("esta llave cincelada")—like the sculpted lines of verse—no longer serves a practical purpose; it has no pragmatic end. By holding on to it, however, the poet asserts the profound "usefulness" of its "uselessness": the key is able to conjure up those things that have disappeared from sight, to recall the past, to evoke memories, to, in short, put its owner in touch with intangible truths. For this reason, Nervo writes: "Me parece un amuleto / sin virtud y sin respeto . . ." ["It seems to me an amulet / without virtue or respect . . ."]. Judged harshly by a world that holds dear only those things that "work," the key, like the poem itself, opens doors to a magical realm that is overlooked by those who live by the values of modern society. By attaching these powers to an object of sentimental value—an object that evokes biographical, religious, and national emotions—Nervo concretizes an abstraction and affirms the multifaceted mission of *modernista* discourse.[5]

The political dimension alluded to in "Vieja llave" takes center stage in works like *La raza muerta* [*The Dead Race*] (1896) (2:1352–1353); "Guadalupe la Chinaca" ["Guadalupe the Destitute"] (1899) (2:1489–1491); "La raza de bronce" ["The Bronze Race"] and "Canto a Morelos" ["Song to Morelos"] (from *Lira heroica* [*Heroic Poetry*], 1902) (2:1403–1409, 1410–1415); and "Los niños mártires de Chapultepec . . ." ["The boy martyrs of Chapultepec . . ."] (1903) (2:1492). As José Emilio Pacheco has noted, with these pieces Nervo anticipates the Chocano of *Alma América* [*Soul America*] (1906), the Darío of *Canto a la Argentina* (1914), and the Lugones of *Odas seculares* (1910) as he attempts to strike a balance between cosmopolitanism and patriotic concerns (2:4). This focus on national issues appears throughout his extensive critical writings as well. In these prose pieces, he refers to literary enterprises as a possible antidote to the prevalent stultification of the young and their unbridled regard for money.[6] He also defends, as noted in chapter 2, *modernista* innovations in poetic form and language as a necessary and enlightened adjustment to the evolving social context within Mexico. With this attention to national concerns, his nostalgic vision of provincial life, his eroticism tinged with guilt, together with his search for a philosophy of life, Nervo lays the groundwork for later poets, most notably his compatriot Ramón López Velarde.

Nervo pursues these themes throughout the three periods into which his work can be divided. The first includes *Místicas* [*Studies of the Contemplative Life*] (1898), *Perlas negras* [*Black Pearls*] (1898), *Poemas* [*Poems*] (1901), *La hermana agua* [*Sister Water*] (1901?), *El éxodo y las flores del camino* [*The Exodus and Flowers along the Road*] (1902), and ends with the political *Lira heroica* [*Heroic Poetry*] (1902). In these collections Nervo confesses his personal obsession with erotic passion and religious doubt. He finds consolation in the stimulatingly unorthodox ideas offered by esoteric thought, which surfaces in the pantheism of *La hermana agua* and in the figure of the cosmic androgyne in "Andrógino" ["Androgyne"] (*Poemas*) (2:1349).[7]

The second stage consists of *Los jardines interiores* [*Inner Gardens*] (1905) and *En voz baja* (1909), the collection in which "Vieja llave" appeared. As Merlin Forster notes, Nervo's poetry at this point becomes more serene. He finds a way of accommodating his spiritual concerns and anxieties within the context of the material—and materialistic—world in which he lives (86). In "Mi verso" ["My Poetry"], the second poem from *Los jardines interiores,* Nervo presents an *ars poetica* in which this accommodation predominates. He sets out to forge verses whose worth is evident to all readers, even to those only interested in royal splendor. In this way, he seeks to create a place for himself within a society that would tend to ignore his contributions.

Querría que mi verso, de guijarro,
en gema se trocase y en joyero;
que fuera entre mis manos como el barro
en la mano genial del alfarero.
Que lo mismo que el barro, que a los fines
del artífice pliega sus arcillas,
fuese cáliz de amor en los festines
y lámpara de aceite en las capillas.
Que, dócil a mi afán, tomase todas
las formas que mi numen ha soñado,
siendo alianza en el rito de las bodas,
pastoral en el índex del prelado;
. . . .
Yo trabajo, mi fe no se mitiga,
y, troquelando estrofas con mi sello,
un verso acuñaré del que se diga:
Tu verso es como el oro sin la liga:
radiante, dúctil, poliforme y bello. (2:1533–1534)

[I would want my poetry, made of stone, / to be turned into gem and jewel-box, / to be in my hands like clay / in the ingenious hand of the potter.

That the same as mud, which to his ends / the artist molds into potter's clay, / it should be a chalice of love in the feasts / and oil lamp in the chapels. /

Docile to my anxiety, it should take all / forms that my genius has dreamed, / being bond in the ritual of marriage, / pastoral poem on the index of the prelate; / . . .

I work, my faith does not diminish, / and minting stanzas with my seal, / I will coin a verse of which is said: / Your verse is like gold without the alloy: / radiant, malleable, diverse in nature, and beautiful.]

The central comparisons of potter and poet and of clay and language dramatically underscore Nervo's goals and vision. Not only does he wish to transform the inchoate and unserviceable raw material of communication into an artistic creation, but he wants his poetry to be versatile, to adapt to all settings and all sentiments. If from clay the potter can produce both chalice and lamp, the poet similarly aspires to provide—through his art—both an elixir of love and spiritual illumination, the impact of which he hopes will extend beyond the "chapel," into the rites and routines of daily life. For his poems to achieve this diversity, they must respond to his spirit, they must accommodate his dreams. With this imagery, Nervo reiterates the search for a language that is fluid and responsive rather than preset or premolded. The word chosen in the last line is "dúctil" ["malleable"]. Significantly, however, by the final stanza the image has changed. The comparison is not with clay but with gold and the closing simile picks up the regal tone of the initial two lines of verse. Nervo's desire is ultimately to convert the substance of everyday reality into objects of transcendent beauty and worth that are also coveted by the society at large. If Nervo wants to "mint" stanzas and turn verses into gold, his goal is to be able to move from what he called the "aristocracia en harapos" ["aristocracy in rags"] to the aristocracy of money and power from which he felt excluded but to which he felt—by privilege of merit—he belonged. The poet's aspiration to wealth—even if only verbal—is exemplary of his conflictive attitudes, not uncommon among *modernistas,* toward the dominant values of the day and the materialistic society that fostered them.[8]

The third period of Nervo's poetry was colored by the death of his beloved wife Ana Cecilia Luisa Dailliez in 1912 and by a search for consolation that never fully relieved his unremitting grief. Many of his most tortured poems were written in the year of her death and appear in *La amada inmóvil: Versos a una muerta* [*The Constant Beloved: Poetry to a Dead Woman*], which was not published until 1920. The later collections, *Serenidad* (1914), *Elevación* [*Exaltation*] (1917), *Plenitud* [*Plenitude*] (1918), *El estanque de los lotos* [*The Lotus Pool*] (1919), and *El arquero divino* [*The Divine Archer*] (1920), explore the possibilities for solace and transcendence offered by both Christian and oriental philosophies.

Other nontraditional perspectives were influential in the works of Bolivian-born Ricardo Jaimes Freyre, which continue to be read and studied today for three important reasons: (1) his active collaboration in Buenos Aires with Darío, with whom he founded in 1894 the short-lived but influential *Revista de América*, (2) his innovative adaptation and enthusiastic defense of free verse,[9] and (3) his syncretic recourse to medieval and Nordic myths and legends to express concerns regarding the general artistic and sociopolitical context in which he wrote.

Jaimes Freyre's attention to the possibilities of versification and the innovation of poetic form is related to his recourse to medieval and Nordic myths and legends. Through both features, he struggles to achieve a poetic vision that is in touch with the primordial rhythms of existence and the sense of a simpler time and place. This perspective is offered as an alternative to the complications and disappointments of modern life. Ricardo Gullón acknowledges this connection. He regards Jaimes Freyre's "exoticism," like that of other *modernistas*, to be an act of "rebellion started in contact with wretched reality. . . . a rebellion against the destiny of man, not only condemned to die, but also to live in societies ruled by the crudest materialism" (279).[10] For Gullón "it is the same to take refuge in the foggy mist of the North—like Ricardo Jaimes Freyre—in the Versailles of Darío, or in the orientalisms of Julián del Casal" (295).[11] Yet, while Jaimes Freyre's unique response to modern conditions is comparable to that of his *modernista* contemporaries, the myths and legends that he turned to evoke different associations than those traditionally linked to *modernismo* or to Spanish America. The strongest associations are with the composer and artistic innovator Richard Wagner, through whose work Jaimes Freyre probably first came to know the unfamiliar northern landscapes and to recognize how these images and themes might serve to articulate his own political and aesthetic concerns.[12]

Certainly Jaimes Freyre was not the only *modernista* to sense this affinity. In two early, uncollected sonnets grouped together under the title "Wagneriana," Darío makes reference to two of Wagner's operas.[13] In "Palabras liminares" (*Poesía* 179) and in "El cisne" ["The Swan"] he goes a step further and explores the role that Wagner might play for *modernista* writers. In the second of these two pieces, Darío presents Wagner as an artistic mentor.

> Fue en una hora divina para el género humano.
> El Cisne antes cantaba sólo para morir.
> Cuando se oyó el acento del Cisne wagneriano
> fue en medio de una aurora, fue para revivir. (*Poesía* 213)

[It was at a divine moment for the human species. / Before the Swan only sang to die. / When the accent of the Wagnerian Swan was heard / it was in the middle of a sunrise, it was to live again.]

The rebirth alluded to in Darío's poem is made possible through "la nueva Poesía" ["the new Poetry"], which brings forth eternal perfection and purity. In this context, Wagner is converted into the epitome of the artist whose work is able to usher in a new day for all of humanity.

David C. Large and William Weber explain how and why Wagner's push for aesthetic and social change attracted so many followers:

> . . . in calling for reform of musical theater he raised a set of issues that on a far broader plane mattered greatly to the people of his time. When he called for opera to function not as entertainment but as the most serious form of art, he rallied the idealists from all the other arts. When he let loose polemics against the frivolous operagoing public, he awakened a widespread resentment against the urban upper classes. When he declared that his theater would reunite the *Volk* and revive the sense of its past, he offered a common meeting ground to diverse cultural and political groups searching for links with popular traditions. When he dwelt upon the psychological and religious themes of his music dramas, he opened up an exciting new direction to people then turning away from positivism and utilitarian rationalism. And when he inveighed against crass materialism and held up love—both spiritual and sensual—as the best antidote to mammon, he appealed not only to the professional bourgeoisie baiters among the intelligentsia, but also to those

important elements among the middle classes given to ostentatious self-laceration and public declarations of their sins against culture. (20)

Exactly which features might have drawn Jaimes Freyre toward the Wagnerian "barbaric Castalia" remain unclear. What this brief assessment makes evident, however, is that Wagner's situation paralleled that of *modernista* authors and that his masterpieces provided original models for responding to the dilemmas that they faced in their own countries. Like the German composer, whose incursions into his pre-Christian Walhalla reflected hopes for political change and a search for redemption from the crass and sterile realities of nineteenth-century Germany, Jaimes Freyre, with his Germanic gods, heroes, and genies, offers an alternative vision to his immediate milieu, one in which the world is alive with spirits, vibrant with song, and saturated with the lifeblood of existence. While the title of this first collection recalls Leconte de Lisle's *Poèmes barbares* (1854) and points to a source of inspiration other than the *castalia clásica* [classic Castalia] of traditional Spanish poetry, the freedom of the verse and the energy of the subject suggest a desire to tap a primitive—not Parnassian—power and to evoke with symbolist magic and musicality an uncorrupted energy that underlies all life and action.

Following the example of other *modernistas* as well as of Wagner himself, Jaimes Freyre relies on a syncretic blending of pagan and Christian symbols and motifs throughout his poetry.[14] Perhaps the best-known example is "Aeternum vale," in which the world order envisioned in Germanic mythology gives way to another based on "el Dios silencioso que tiene los brazos abiertos" ["the silent God whose arms are open"], in what is usually assumed to be a redemptive embrace.[15] Yet the symbolism is ambiguous. The order that comes to an end is described in the last two stanzas—as throughout—in adulatory terms.

> Ya en la selva sagrada no se oyen las viejas salmodias,
> ni la voz amorosa de Freya cantando a lo lejos;
> agonizan los Dioses que pueblan la selva sagrada,
> y en la lengua de Orga se extinguen los divinos versos.
> Solo, erguido a la sombra de un árbol,
> hay un Dios silencioso que tiene los brazos abiertos. (30)

[The old psalmodies are no longer heard in the sacred forest, / nor Freya's loving voice singing in the distance; / the Gods that inhabit the sacred forest are dying, / and in the Orga language the divine verses are extinguished. /

Alone, erect in the shade of a tree, / there is a silent God who has his arms open.]

The penultimate stanza expresses a profound sense of loss. The harmony felt within the "sacred forest" disappears. The loving music and divine poetry are replaced by absolute quiet, and even though we may assume that the silent God is Jesus and that his silence evokes his martyrdom, the transition to the new era appears ominous. The nostalgia with which Jaimes Freyre depicts the earlier period, its energy and its spirit, highlights the longing for other times when people could still feel the pulse of the universe. This "Götterdämmerung" ["Twilight of the Gods"], not unlike Wagner's, reflects a disappointment with the unseemly side of life that has robbed the Gods of their strengths and virtues and men of the ability to perceive them. Both the opera and the poem, however, add an optimistic coda by announcing a new chapter in history. In Wagner's masterpiece free men create the new order.[16] In Jaimes Freyre's, Jesus presides over the future.

Jaimes Freyre's second collection of poetry, published eighteen years after the first, is less Germanic and more universal in inspiration. It develops both the philosophical and political aspects implicit in many of the poems of the earlier collection. In the section "Las víctimas" ["The Victims"], the voice of the oppressed is heard denouncing injustice. The position taken by the poem "Clamor" ["Outcry"] (119–123) anticipates the rewriting of elite history that emerges in *Canto general* over thirty years later. Similarly, "Rusia" ["Russia"] (129–130), written in 1906, demonstrates both a profound sympathy for the masses and a prophetic sensibility regarding the possibility of social upheaval. The politics that appears in his writings was not, however, an occasional intellectual exercise for Jaimes Freyre. Once back in Bolivia, after having taught for many years in Tucumán, Argentina, he assumed several high administrative posts and even considered running for office.

Even more than for Jaimes Freyre, politics was a central activity in the life of Guillermo Valencia. He served in the Colombian Congress, in high administrative positions, and in the diplomatic corps. He was also twice named the Conservative Party candidate for president—though he was never elected to the office. Despite his public service, he maintained an active literary career, a large part of which revolved around translation. His sense of good taste, his solid humanist education, and his knowledge of several classic and modern languages are evident throughout his sensitive and skilled translations, the earliest of which were of his European contemporaries and immediate predecessors—Keats, Hugo, Flaubert, Heine, Baudelaire, Gautier, Leconte de Lisle, Verlaine, D'Annunzio, Stefan George, Hugo von Hofmannsthal, and Oscar Wilde—as well as of the Indian writer and philosopher Sir Rabindranath Tagore.[17] Later, in *Catay* (1929), he presented idiosyncratic versions of Chinese poets such as Li-Tai-Po, Tu-Fu, and Wang Hei based on the French prose translations by Franz Toussaint. The minimalist nature of these oriental pieces reinforced poetic trends developing throughout Spanish America at the time. During the same period, José Juan Tablada (Mexico, 1871–1945), who is generally classified as a *posmodernista* poet, was introducing the elegant simplicity of Japanese haiku to Spanish.

Valencia's own poetry consists of one book of verse, *Ritos* [*Rituals*], which was originally published in 1899 and which reappeared in an expanded version in 1914. This work is dominated by a seriousness and a symbolic density, by a linguistic sophistication, terseness, and polish, that have been characterized both as classical and as Parnassian. But his work, which also shows a sensual and symbolist attention to the tone and texture of verse, defies easy classification. As has often been noted, Valencia's desire to celebrate "literary rites" counters the negative forces that impede the spiritual elevation of the individual and society. This effort—encased within a poetry of verbal splendor, encyclopedic breadth, and punctilious precision—links Valencia with other *modernistas.*

Equally difficult to classify is the work of José María Eguren. He published his first poems in the magazines of Lima around 1899 and his first book, *Simbólicas* [*Symbolic Poems*], in 1911. With *Simbólicas* he began a poetic career that was generally misunderstood and much maligned. His work contradicted the emphatic and declamatory poetry popular at the time, a poetry epitomized by the work of Peru's soon-to-be poet laureate José Santos Chocano. Although he went on to publish two other collections, *La canción de las figuras* [*The Song of Images*] (1916) and *Poesías: Simbólicas, La canción de las figuras, Sombra, Rondinelas* [*Poems: Symbolic Poems, The Song of Images, Shadow, Rondinelas*] (1929), he remained throughout his career a little-known, marginal poet. His supporters—such illustrious Peruvians as Abraham Valdelomar, Manuel González Prada, José Carlos Mariátegui, and César Vallejo—praised his radical independence, his originality, and his unique and solitary nature but were unable to overcome the widespread public and critical antipathy to and misapprehension of his work.[18]

If Eguren was criticized during his lifetime for being difficult, obscure, and "hermetic," he is now esteemed as the supreme representative of symbolism in Peru and even as a forerunner of the avant-garde. His work—with its references to the night, to the tenuous and frightening realm of childhood memories, and to the world of nature turned unreal—reveals the poet's sense of mystery and awe regarding the order of things. From the outset, Eguren's poetic imagination proposes a form of spiritual idealism with which he counters the burden of personal tragedies and social ills. Focusing on this feature of his work, James Higgins shows that Eguren viewed the society of his day with a critical eye. In opposition to the "negative lifestyle" of the modern world, to its malice and viciousness, and to its spiritual bankruptcy, Eguren provides a competing set of personal values. He presents nature with a wonderment that borders on the religious and conveys this vision through a fantastical language and musicality that, while derivative of *modernista* innovations, is uniquely his own.

Eguren's work was deemed marginal to *modernismo* because of its symbolist tendencies, its lack of epic breadth and tropical lushness, and the absence of Versaillesque gardens and swans.[19] The creative features of his poetry which shocked and alienated his early readers overshadowed all else. Today, however, his neologisms and his scenes of childhood innocence turned menacing clearly resonate with echoes of the fanciful inventions of *modernismo* and its search for the paradise lost to modern man. They also anticipate the loss of faith and the fearful visions of the avant-garde, most notable in the poetry of his fellow Peruvian César Vallejo. Julio Ortega, examining this aspect, points out that in Eguren's poetry innocence tends to slip into horror. He writes: "His friends and critics saw in the childhood scenes of this poetry the guileless poet that appeared to be playing. They did not see another thing: the horror of this game, the look of the poet attacked by fear, life, and death which Eguren wanted to show in all its fragility in a repeated childhood scene" ("José María Eguren" 65).[20]

A poem that exemplifies Eguren's unique poetic amalgam is "El duque" ["The Duke"]. The singsong rhythm and emphatic rhymes evoke childhood doggerels of little consequence. Yet the scene of children's play reverberates with social commentary and the menacing tone of shattered expectations.

> Hoy se casa el duque Nuez;
> viene el chantre, viene el juez
> y con pendones escarlata
> florida cabalgata;
> a la una, a las dos, a las diez;
> que se casa el Duque primor
> con la hija de Clavo de Olor.
> Allí están, con pieles de bisonte,
> los caballos de Lobo del Monte,
> y con ceño triunfante,
> Galo Cetrino, Rodolfo montante.
> Y en la capilla está la bella,
> mas no ha venido el duque tras ella,
> los magnates postradores,
> aduladores
> al suelo el penacho inclinan;
> los corvados, los bisiestos
> dan sus gestos, sus gestos, sus gestos;
> y la turba melenuda
> estornuda, estornuda, estornuda.
> Y a los pórticos y a los espacios
> mira la novia con ardor; . . .
> son sus ojos dos topacios
> de brillor.
> Y hacen fieros ademanes,
> nobles rojos como alacranes;
> concentrando sus resuellos
> grita el más hercúleo de ellos;
> ¿Quién al gran Duque entretiene?
> ¡ya el gran cortejo se irrita! . . .
> Pero el Duque no viene; . . .
> se lo ha comido Paquita. (53–54)

[Today Duke Nut gets married; / the choirmaster is coming, the judge is coming / and with banners the flowery / red cavalcade; / on the one, on the two, on the ten; / Duke excellence is marrying / the daughter of Cloves. / There they are, with buffalo skins, / the horses of Wolf of the Mountain, / and with a triumphant expression, / Galo Yellowcolored, Rodolfo Broadsword. / In the chapel is the beauty, / but the duke has not come behind her, / the flattering / fawning magnates / bow their feathered heads to the floor; / the bent-over and leap-year-added individuals / offer their gestures, their gestures, their gestures; / and the hairy crowd / sneezes, sneezes, sneezes. / And the bride looks at / the halls and spaces with ardor; . . . / her eyes are two topazes / of lustre. / And the noblemen, red like scorpions, / make fierce gestures; / concentrating his breath / the most Herculean of them shouts: / Who detains the great Duke? / the court is already getting irritated! . . . / But the Duke does not come; . . . / Paquita has eaten him all up.]

Nobility, luxury, social climbing, and self-importance are all ridiculed in this childlike imitation of adult behavior. From this perspective the ostentation of the rich and the obsequious behavior of their retinue is humorously absurd. But, even though the protagonists in this marriage are candy replicas of human models, the abrupt ending is disquieting, a not-so-innocent rendition of real-life tragedy. The final line of verse demands that the poem be reread with an awareness that disaster hides behind the festive facade presented.

This distrust of appearances and sense of dislocation expresses a modern dread and anxiety about the precarious nature of a previously taken-for-granted social world whose foundations are being jolted by modernizing processes. Eguren's poetry responds to this changing context. It contains an innovative mix of elements that underscores the fluidity among the key movements of the turn of the century—*modernismo, posmodernismo,* and the early avant-garde, all of which confront the unanticipated ramifications of modern life.

Unlike the work of Eguren, the poetry of José Santos Chocano does not anticipate the future but rather looks back toward the past. It appeared on the scene as an exuberant affirmation of the painterly, elaborate, image-laden *modernismo* that had for the most part begun to recede in favor of a more introverted and intimate perspective or a more radical, volatile style. Perhaps because of this very fact—together with his enthusiastic pro-American *Mundonovismo*—his poetry pleased a wide audience accustomed to *modernista* fare, and he became one of the most popular poets of the time.

Chocano was driven by the idea of becoming "the poet of America" and unself-consciously called up real and imagined images of South America in order to achieve this title—one that had been foreclosed to Darío by José Enrique Rodó's early assessment of his work. In his poetry, South American geography, history, legend, landscapes, peoples, flowers, and animals are painted in broad and vibrant brushstrokes. His natural inclination toward grandiloquence, linguistic inflation, and rhetorical expansiveness found a perfect outlet in his purposeful exaltation of American enterprises and wonders. Chocano would declaim—with great personal and economic success—his poetic depiction of these activities and marvels to audiences in Madrid, the Caribbean, and Central and South America, and the oratorical nature of these recitals further encouraged his established stylistic tendencies. Thanks to his fame, popularity, and the pose he assumed as he executed his poetic goals, he was crowned national poet of Peru in 1922, though he hardly ever lived there. Since the end of *modernismo*, the bombast, hyperbole, excess, and lack of nuance of his work have tended to receive as much commentary as his accomplishments. His reputation as a poet was further hurt by his soldier-of-fortune lifestyle. Not only did he spend a year in jail for killing a young man in 1925, but he himself was killed in 1934 by a Chilean worker who believed that he had been bilked out of money in one of Chocano's many get-rich-quick schemes. Despite the overly theatrical strains in his life and works, it is easy to recognize why his verse was received with such enthusiasm during the forty-five years that encompass the period between the publication of his first collection, *Iras santas* [*Holy Angers*] (1895), and that of his thirteenth (and posthumous), *Oro de Indias* [*Gold of the Indies*] (1941), including his principal work, *Alma América: Poemas indoespañoles* [*Soul America: Indo-Spanish Poems*] (1906). His vision was Hispanic rather than Continental; he even criticized Darío for his recourse to French models and chided other Latin Americans for their fascination with Europe. His pride in America and his hopes for its future—evident, in part, in his vociferous support of the Mexican Revolution—are expressed in a language that is stirringly dramatic, strong, and assertive. The power of his poetry relies upon a keen sense of rhythm and an unflinching sense of purpose. His tone is overwhelmingly optimistic, confident, and direct; philosophic anguish and doubt as well as poetic suggestion and subtlety were left to others. "Blasón" ["Coat of Arms"], from *Alma América,* reveals both his artistic strengths and the egomaniacal nearsightedness that has made him unpalatable to some modern readers.

> Soy el cantor de América autóctono y salvaje:
> mi lira tiene un alma, mi canto un ideal.
> Mi verso no se mece colgado de un ramaje
> con un vaivén pausado de hamaca tropical . . .
> Cuando me siento Inca, le rindo vasallaje
> al Sol, que me da el cetro de su poder real;
> cuando me siento hispano y evoco el Coloniaje,
> parecen mis estrofas trompetas de cristal.
> Mi fantasía viene de un abolengo moro:
> los Andes son de plata, pero el León de oro;
> y las dos castas fundo con épico fragor.
> La sangre es española e incaico es el latido;
> ¡y de no ser Poeta, quizás yo hubiese sido
> un blanco Aventurero o un indio Emperador! (381)

[I am the singer of native and wild America; / my lyre has a soul, my song an ideal. / My poetry does not rock from branches / with a measured sway of a tropical hammock. . . .

When I feel Incan, I render obedience / to the Sun, which gives me the scepter of its royal power; / when I feel Hispanic and I evoke Colonial rule, / my stanzas seem crystal trumpets.

My imagination comes from a Moorish ancestry: / the Andes are of silver, but the Lion is of gold; / and I fuse the two lineages with epic noise.

The blood is Spanish and Incan is the beat; / and if I were not a Poet, perhaps I would have been / a white Adventurer o an Indian Emperor!]

Chocano's poetry echoes the vocabulary and grammar of *modernismo* but does not reflect the tension and anxiety that so energized the movement. The struggle between poet and society recedes, and what was a language of protest and antagonism is hollowed out. Chocano's poetry fuses with the discourse of power and becomes part of the "official" language of Spanish America. Yet, for *modernismo* to keep on course and true to its original program, it had to maintain its critical stance. It had to persist in stepping beyond the routine; it had to persevere in its confrontation with predominant values. In short, it had to maintain the gap between "rhetoric" and "literature."[21] The three poets discussed in the next chapter assert this antagonistic relationship with society and continue to revitalize the expression of epistemological and political concerns.

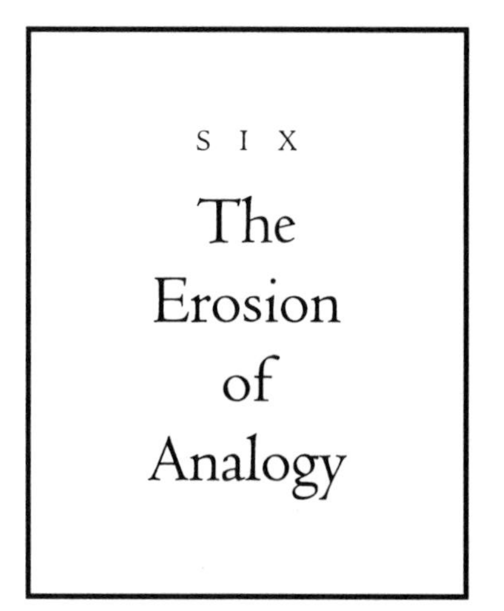

SIX

The Erosion of Analogy

This chapter centers on three poets whose works signal the start of the transition from *modernismo* to the avant-garde. They are Leopoldo Lugones, Julio Herrera y Reissig, and Delmira Agustini. Their writings reveal a profound, almost tragic realization which Octavio Paz, in *Los hijos del limo,* associates with the onset of the avant-garde. Paz holds that the avant-garde begins when the belief in analogy erodes and dissonance takes over, and he identifies this breakdown with irony in poetry and with mortality in life. Much earlier, in his famous essay on Lugones, Borges had focused on the same fundamental shift in worldview; he describes the poet as "a man who controlled his passions and industriously built tall and illustrious verbal edifices until the cold and the loneliness got to him. Then, that man, master of all the words and all the splendor of the word, felt within his being that reality is not verbal and may be incommunicable and terrible, and went, silently and alone, to look, in an island's twilight, for death" (97).[1]

Borges refers here both to the poet's suicide and to the loss of faith in the decipherability of the universe, a loss that marks an essential realignment within *modernista* poetry. Precedents for this stance, however, can be found in the movement itself. The fear that language would not give up its secrets and that the universe would remain a mystery, at times a cruel and unforgiving mystery, is evident in Silva's *Gotas amargas,* in Darío's poems of despair and tortured self-doubt, and in Eguren's scenes of innocence turned menacing. Nevertheless, despite occasional assertions to the contrary, *modernismo*'s overriding faith is that poetry will ultimately provide a magic code capable of revealing the beauty of creation and the meaning of existence that, in itself, is a commentary on the stranglehold of everyday life and those caught up in it.

Early in the movement, the perfection of the universe was linked to the beauty of art as well as to opulent elegance. The enthusiastic embrace of elements of high culture from all corners of the world (including Spanish America) gave *modernismo* what is, for contemporary readers, a fantastical, even cloying, richness.[2] Perhaps because art and artifice became associated with artificiality and even insincerity (just the opposite of what the *modernistas* aspired to), succeeding generations of readers failed to see in these elaborate images and structures the complex critique of their life and times that *modernista* writers were making. They failed to understand that *modernistas* believed that the truth of poetry was revealed in—not obscured by—ornamentation. As *modernista* discourse matured and acquired a sense of self-worth, poetic expression became simpler, more direct. The language of *modernismo* made the transition championed by González Martínez in his demand for change. At the same time, however, some *modernistas* began to express doubts about the power of poetic discourse to fill the voids left by the disruptions of modern life. As this pessimism moves from periphery to center, the discursive adjustment underscored by Paz occurs: irony peeks through the controlled and crafted surface of *modernista* verse. This development begins in earnest with Lugones and Herrera y Reissig and continues decisively into the avant-garde.

Paradoxical as it may seem, this ironic and irrational use of language is yet another outgrowth of *modernista* discourse. *Modernistas* had rejected a rigid sense of referentiality and had turned away from "realism." They explored numerous symbolic systems, always seeking a fundamental unity and harmony among them. The fires of this aspiration were fueled by the occultist sects that had become extremely popular at the time. This syncretism—exceptionally strong in the works of Darío, Nervo, and Jaimes Freyre but prevalent virtually throughout the movement—pulled symbols from their original codes and juxtaposed them with others. The interplay of disparate referential structures, including those of everyday reality, permitted a new conception of language and representation. No longer fixed, signs and symbols (once they failed to reveal the much pursued harmony of existence) began to collide, setting up a chain reaction that would lead to the free play of the avant-garde and eventually to the deconstructionist discourse of today's postmodernism.

As the poetry of Leopoldo Lugones reveals, however, the trajectory just outlined is far from linear. Lugones's writings include texts that are both radically innovative and nostalgically conservative. For this reason, it has virtually become a truism of *modernista* criticism to say that there is no one Lugones but rather several. While the diverse nature of Lugones's poetic production has long been recognized, only recently, with studies such as those by Saúl Yurkievich and Gwen Kirkpatrick, has the basis for this shifting pattern of imitation and innovation begun to be recognized. The contradictions and asymmetries within Lugones's poetry reflect a development within *modernista* poetics that arises from a changing artistic and sociopolitical context and that eventually leads to a distancing from and even a disenchantment with the poetic vision of the movement—a disenchantment that anticipates the poetic transformations of the avant-garde. In other words, the dislocations within the cultural frameworks of a modernizing Argentina provide the backdrop to Lugones's many poetic and political voices.[3]

Either in spite of or because of these fundamental shifts and the cataclysmic upheavals that they generated, Lugones rather quickly came to turn his back on innovation and the unknown and chose to affirm the great traditions of rhyme and patriotism. Starting as early as 1910 with *Odas seculares*, Lugones's experimentation recedes and conservative values and visions begin to take over. Until then, in his first three collections, *Las montañas del oro* [*The Mountains of Gold*] (1897), *Los crepúsculos del jardín* [*Garden Twilights*] (1905), and *Lunario sentimental* [*Sentimental Lunar Calendar*] (1909), the impetus is toward the deliberately new.

For this reason, Kirkpatrick places Lugones at the center of the initial move away from *modernismo.* Though her enthusiasm for Lugones is at times excessive, her conclusions are worthy of note:

> The tear Lugones made in *modernismo*'s fabric of social and sexual dynamics is still being rewoven by contemporary poets. Lugones' intrusiveness created a lingering discordance, and no amount of dispassionate criticism can gloss over the uneasy spaces he created. Julio Herrera y Reissig, César Vallejo, Ramón López Velarde, Alfonsina Storni, to mention a few, are poets who have not let us forget

> this rupture. Marked by violence, eroticism, and the disturbing entrance of urban elements in a textual space, these poets struggle with an ambivalence against allowing easily mappable patterns of perspective, beauty, and poetic structure to frame their poetry. The subversive shifts and overt disavowals they make of a veiled authoritative order are the weapons they use in dismantling hierarchical forms, including a realignment of the speaking subject. They are not simply naive consumers of European influences. Each in his own way plots a path to lead the reader to question even the poetic forms that tradition supplies. (*Dissonant Legacy* 11)

Lugones's first collection, *Las montañas del oro* (1897) represents both a break and a continuation, for it both embodies and exaggerates the *modernista* delight in presenting the unexpected and unorthodox and in engaging in the iconoclastic playfulness that unsettled readers of, for example, Darío's *Prosas profanas.* In this collection, Lugones placed, next to prose pieces, poems in which the verses run on, separated only by hyphens, and he purposely replaced in them the *y* with *i,* an orthographic change that later editions do not always respect. But Lugones's contributions went beyond willful alterations in poetic form. He was able to successfully incorporate into this work the prophetic splendor of Hugo, the resonant serenity of Whitman, the visionary intensity of Dante, and the fundamental power of Homer. In the first poem, which is simply called "Introducción" ["Introduction"], Lugones explicitly aligns himself with these great poets of the past and presents a poetic program which, for the most part, coincides with the premises of the *modernista* verse being written at the time. Consistent with *modernista* beliefs, he lays claim to a special relationship between the poet and God: the poet hears the voice of God in nature and creates a spiritual force in his verse that moves the earth and lights the darkness of despair and ignorance. He ends "Introducción" by singling himself out as the one to "read" the signs of the universe:

> Nadie alzaba los ojos para mirar aquellas
> Gigantes convulsiones de las locas estrellas;
> Nadie le [*sic*] preguntaba su divino secreto;
> Nadie urdía la clave de su largo alfabeto;
> Nadie seguía el curso sangriento de sus rastros
> Y decidí ponerme de parte de los astros. (60)

[No one raised his eyes to look at those / giant convulsions of the crazy stars; / no one asked them their divine secret; / no one twisted the key of their long alphabet; / no one followed the bloody course of their trails

And I decided to align myself with the stars.]

The poet feels himself alone in his quest for knowledge, knowledge which is both worldly and transcendental. These two facets come together in his warnings against the Herculean strength of "tío Sam" ["Uncle Sam"] (58), against whose power "clouds and dawns" should be marshaled in support. With these alternate resources at his disposal, he optimistically proclaims the New World "the great reserve of the future" (59).

The revitalizing impetus revealed in "Introducción" is broadened throughout the collection. By the final pieces of the work's third cycle, it encompasses a dizzying array of scientists, theosophists, philosophers, economists, and artists. This syncretic blending of ancient occult and modern physical sciences, together with the fusion of the arts and philosophy, is offered in a frenzied embrace of progress similar to the Futurism proposed by Marinetti twelve years later but humanized and spiritualized by a faith in a divine and harmonious order.[4] The overall impression, however, remains consonant with Darío's early assessment of Lugones's poetic grasp of the world around him as "un rápido choque de miradas" ["a rapid collision of glances"].[5] The sense of excessive accumulation and, occasionally, transgressive defiance will continue in different guises and to different degrees in the next two collections.

The publication of *Los crepúsculos del jardín* in 1905 builds upon *modernismo*'s foundations in symbolism and the latter's emphasis upon correspondences between the visible and transcendent realms of existence. The strongest element in this collection is the visual. The depiction of garden scenes, of patterns of fading light, and of erotic encounters evokes subtle dimensions of reality, many of which are menacing and disquieting. This pictorial representation and the hidden forces it reveals link Lugones with French symbolist Albert Samain, whose *Au Jardin de l'infante* (1893) is considered to have been a direct influence on both *Los crepúsculos del jardín* and on Herrera y Reissig's *Los éxtasis de la montaña* [*The Ecstasies of the Mountain*], a confluence of inspiration that led to an unfounded accusation of plagiarism against Lugones in 1912. Along with Samain, the influence of Verlaine and Darío is also evident in this collection. Unlike these poets, however, Lugones appears reluctant to maintain the sense of mystery and musicality that defines the symbolist substructure of his work. He often interrupts the mood created with an unexpected term, incongruous image, or transgressive allusion. The effect can be parodic, comic, decadent, or—as in "El solterón" ["The Old Bachelor"] and "Emoción aldeana" ["Rustic Emotion"]—a strange cross between ironic, nostalgic, and restrained.[6]

The most disturbingly moving poems in *Los crepúsculos del jardín* are the twelve sonnets that make up "Los doce gozos" ["The Twelve Pleasures"] (117–124). These sonnets develop the eroticism and violence of the first two cycles of *Las montañas del oro.* Unlike the fanciful, playful, or irreverent sexuality of Darío's *Prosas profanas,* they reflect the dark side of passion, still capable of revealing what is hidden from view by good manners and social constraints but frightening in their evocation of destructive forces. Instead of building a vision of peace and fulfillment, these poems conjure up the decadent pleasures of willfully stepping beyond accepted limits and embracing the underside of existence. This move toward an avant-garde perspective emerges within a recognizably *modernista* framework. For example, in "Oceánida" ["Ocean Nymph"], Lugones plays with the promise of harmony. He draws upon familiar *modernista* images of the lapping of the waves, of an erotically charged universe, and of human reconciliation with the pulse of creation. By associating the rhythmic movement of the ocean with male sexual urges, he eroticizes nature and endorses, as part of the natural order, an integration with these impulses—not unlike some poems from Darío's *Prosas profanas* or *Cantos de vida y esperanza,* which was published during the same year. Lugones writes:

> El mar, lleno de urgencias masculinas,
> Bramaba alrededor de tu cintura,
> Y como brazo colosal, la oscura
> Ribera te amparaba. En tus retinas,
> Y en tus cabellos, y en tu astral blancura,
> Rieló con decadencias opalinas
> Esa luz de las tardes mortecinas
> Que en el agua pacífica perdura.
> Palpitando a los ritmos de tu seno,
> Hinchóse en una ola el mar sereno;
> Para hundirte en sus vértigos felinos
> Su voz te dijo una caricia vaga,
> Y al penetrar entre los muslos finos,
> La onda se aguzó como una daga. (122)

[The sea, full of male urgencies, / roared around your waist, / and like a colossal arm, the dark / shore protected you. In your retinas, and in your hair, and in your astral whiteness, / that light from dying afternoons / that survives on the calm waters / shimmered with opalescent fadings.

Beating at the rhythm of your breast, / the serene sea swelled into a wave; / in order to submerge you in its feline vertigos

its voice told you a vague caress, and upon penetrating between your slender thighs, / the wave became sharp like a dagger.]

At first the poet appears to perceive a universal order that offers security and salvation. The shore provides a protective arm and the afternoon light survives beyond its timely end on the calm waters of the ocean. As the serene sea beats in unison with the breast of the swimmer, perfect accord is achieved. This scene, however, turns violent. The sexually swollen wave becomes a dagger that penetrates the female of the poem. The *modernista* expectation that nature will provide a harmonious alternative to the chaotic confusion of modern life (as it did, for example, in Gutiérrez Nájera's "Para entonces") is turned on its head. The vision of harmony is rent apart by a world that reflects—rather than counters—human aggression. The poem thus reveals truths that are more cautionary than consoling.

Similarly, "Holocausto" ["Holocaust"] suggests a loss of innocence that goes beyond the sexual awakening alluded to in the poem.[7] The grace of the evening landscape is interrupted by the strident cart that returns from a journey described as "espectral" ["ghostly"]. The countryside is thus charged with supernatural possibilities. Seen from the lover's point of view ("a través de tus pestañas" ["through your eyelashes"]), all of nature is imbued with a sense of foreboding—the wind feels contrition, the willows are mournful, day ends in a slow agony and takes on a universal ("astral") melancholy. But this is not a hymn to lost virginity and regret, in which nature echoes the couple's guilt. Quite the contrary, their love "burns like incense in a peaceful combustion of aromas." Their love offers the form of mystical experience typical of *modernista* poems. It is "la sombra pecadora" ["the sinful shadow"] that sees, in the speaker's soul, the wandering off of lambs and, upon the lover's breast, the beheading of doves. Whether the shadow represents conservative moral injunctions that haunt the lovers or a sinister force that emerges unexpectedly, it serves to project an image of love disrupted by discord. The "extravío de corderos" ["the wandering off of lambs"] underscores the poet's loss of direction and confidence, the inversion of the order he expected to find. The brutality of the final image, the beheading of doves, reflects back upon the beautifully constructed sonnet in a gesture that appears to mock the search for perfection and transcendence. The poet has begun to confront a world beyond redemption.

The sonnets comprising "Los doce gozos" thus reveal Lugones's contradictory position with regard to *modernista* poetics. The visionary stance is assumed and the promise of salvation is made, but the universe that is depicted is neither hospitable nor classically beautiful. While their decadent tone had already surfaced in *modernista* verse, in works by Casal and Silva, these poems present more far-reaching and frightening implications, for neither art nor other highly refined activities prove capable of providing a spiritual or emotional haven. In "Delectación morosa" ["Slow Delight"], another sonnet in this grouping, the world itself is askew.[8] With references to bloodless knees upon a pedestal base and a warped sky, "Delectación morosa" suggests a mysterious, almost demonic rite of initiation at the same time that it makes clear that what the poet has come to know has begun to alter the *modernista* worldview. In these sonnets, Lugones faces this fearsome vision with an intriguing and unnerving combination of revelation and revolution.

It is the revolutionary aspect of his verse that comes to the fore in his next collection, *Lunario sentimental,* written under the influence of Jules Laforgue's *L'Imitation de Notre Dame la Lune* and its lyrical irony. In this work Lugones combines prose and poetry, breaks with the affectations of poetic language, introduces colloquial discourse, builds daring metaphors, incorporates unusual adjectivization, dissolves organizing frameworks, and relies upon rhetorical games that underscore his attack against traditional views of poetic beauty. He supports in his prologue the freedom provided by free verse at the same time that he insists upon rhyme as the "elemento esencial del verso moderno" ["the essential element of modern verse"]. But the rhymes that appear are often as disturbingly unexpected as the images. While the volume is unified by the presence of the moon, this sacred icon of poetic discourse, along with other conventional images, is trivialized, caricaturized, and parodied to the point that, as Borges pointed out, the verbal structure becomes the focus of attention much more than the scene or emotions described (33).

These formal changes and, most especially, the freeing of poetic signs from previous constraints, reflect Lugones's changing attitude toward the role of the poet, one that in its own way pushes *modernismo* further toward the avant-garde. As Kirkpatrick has noted, in *Las montañas del oro* the poet, whether as prophetic leader of humanity or as satanic visionary, functions as interpreter of universal meaning. In *Los crepúsculos del jardín* he becomes an organizing voice, rearranging the chaotic welter of elements of the perceptible realm of existence. In *Lunario sentimental* Lugones turns previous conceptions of the poet's role on their heads and expresses the inherent falseness of all things (150). This emotional and conceptual estrangement from the taken-for-granted encourages artifice, comedy, and a disconcerting mix of perspective, tone, and language, all of which is offered as a defiant response to a world of appearances.

In a reasoned explanation for the innovative path that he pursues, Lugones offers a practical and patriotic defense of poetry in his prologue to *Lunario sentimental.* He asserts the political undercurrent of the *modernista* agenda and proposes to demonstrate "the utility of poetry in the enhancement of languages" (191).[9] One of poetry's great advantages is its concision: "Being concise and clear, it tends to be definite, adding to the language a new proverbial expression or set phrase that saves time and effort" (191).[10] From this perspective, the power of poetic language is held up as a national asset, one that must be cultivated and cared for. "Language is a social resource, perhaps the most consistent element of national manners and customs" (192).[11] Thus it is evident that, despite the verbal artifice and concomitant iconoclasm of his first three collections, Lugones never loses sight of the need for a modern form of discourse appropriate to the emerging nation-states of Spanish America. This commitment to create a language consonant with Spanish American conditions is found throughout his poetry; it is present in what Yurkievich calls his "neo-realist" poems (*Celebración* 62), in the insertion of shockingly prosaic vocabulary such as "cold cream," "alkaline," "hydraulic," and "sportswoman," and in cacophonous rhymes such as "flacucha / trucha," "dieciocho / bizcocho," "botella / doncella," "fotográfico / seráfico." Consequently, Lugones continues the *modernista* pursuit of a language of national identity, and asserts the power of poetic discourse in the face of both reactionary and modernizing forces.

The collection's "Luna ciudadana" ["City Moon"] (286–289) addresses these linguistic concerns as it portrays the prosaic realities of modern urban life. In the struggle between poetry and the mundane, between the extraordinary and the routine, between hopes and reality, between the moon and the city, it is now the unforgiving ordinariness of daily existence that wins out. A young man, who is referred to as "Fulano" ["so and so"], uses his imagination to lessen the burden of his daily trolley ride through a poor district in the city's outskirts. The young man has neither a unique identity nor a unique vision. When a young woman becomes the object of his thoughts, he dreams the dreams of others and thinks about her in platitudes—related to her name, to what she smells like, to angels of destiny, and to widowed mothers. He can offer no special insight into who she is or what she is about because he is imprisoned by prevailing cognitive frameworks. Nor does he break normative expectations by approaching her. Their encounter is over before it begins, and it fails to interrupt the daily routine by which he lives. His customary meal, a bottle of wine or liquor and roast beef, is his next concern. Habitual patterns take over and make thought a noteworthy event ("Se ha puesto a pensar—¡qué bueno!—en una estrella" ["He has begun to think—how wonderful!—about a star"]).

Muy luego, ante su botella
Y su rosbif, el joven pasajero
Se ha puesto a pensar—¡qué bueno!—en una estrella.
Cuando, de pronto, un organillo callejero
Viene a entristecerle la vida,
Trayéndole en una romanza
El recuerdo de la desconocida.
¡Ah! ¿Por qué no le ofreció una mano comedida?
¿Por qué olvidamos así la buena crianza? . . .
¡Cómo se sentiría de noble en su presencia!
¡Con qué bienestar de hermanos
Comentarían fielmente sus manos
Una hora mutua de benevolencia! (288)

[Later, before his bottle / and his roast beef, the young passenger / has begun to think—how wonderful!—about a star. / When, suddenly, a street-organ / comes to sadden his life, / bringing him in a melody / the memory of the unknown woman. / Ah! Why did he not offer her an obliging hand? / Why do we forget in this way our good upbringing?

How noble he would feel in her presence! / With what fraternal well-being / would their hands comment loyally / an hour of mutual benevolence!]

Even the art with which Fulano comes into contact—first in the form of a street-organ melody and later in the guise of the operatic aria *La donna é mobile*—does not offer a path to transcendence. On the contrary, in the life of this Everyman, art functions like one more conceptual mold that prevents him from understanding his own circumstances and that impedes him from acting on the basis of an informed decision. Fulano is "vagamente poeta" ["vaguely a poet"], with the emphasis on vague. The complexities of modern urban life, with the endless trolley rides through repetitious landscapes and hateful restaurant meals, all conspire to drum the originality, the individuality, and the sensitivity out of his soul. Lugones makes clear just how far every *fulano* is from the moon and the magic it once held.[12]

Lugones turns his back on this ironic distance, a key element of *vanguardista* poetry, in his next collections. He sets out instead to elaborate on what is only alluded to in "Luna ciudadana," namely, a vision of the new Argentina. National identity becomes the focus of Lugones's next major work, *Odas seculares,* published in 1910 as part of the centennial celebration of Argentina's independence. The collection is divided into four parts: the first consists of a single poem, "A la Patria" ["To the Homeland"]; the second is entitled "Las cosas útiles y magníficas" ["Useful and Magnificent Things"]; the third focuses on "Las ciudades" ["The Cities"]; and the fourth deals with "Los hombres" ["The Men"], including the most typical of all Argentineans, the gaucho. With these poems Lugones confronts issues pertaining to history, geography, and national spirit. With them he begins his journey to a simpler, more direct style, one that distances him from the features of the avant-garde that define his first three collections.

His next three volumes of verse, *Libro fiel* [*Faithful Book*] (1912), *El libro de los paisajes* [*The Book of Landscapes*] (1917), and *Las horas doradas* [*The Golden Hours*] (1922), also seem to respond to the same impetus toward traditional poetic form, intimacy, and simplicity. The first is the most strongly confessional, embracing the delights as well as the despairs of passionate love. The second and third focus more on the exterior world and present both landscapes and miniatures. His final three collections, *Romancero* [*Book of Ballads*] (1924), *Poemas solariegos* [*Ancestral Poems*] (1928), and *Romances de Río Seco* [*Ballads of Río Seco*], published posthumously in 1938, continue the dual exploration of human intimacy and national identity. *Romancero* bears the influence of Heine in its title, its section "Los trece lieders" ["The Thirteen Lieder"], its opening poem, "Gaya ciencia" ["Gay Science"], and in its overall quietly intense, romantic tone. *Poemas solariegos* and *Romance de Río Seco* find the essence of Argentina in the simple elements of life, which are captured either in rich, detailed description or minimalist minipoems. These works, while a step back from the verbal pyrotechnics of the first three collections, continue Lugones's lifelong affirmation of the *modernista* goal of finding a language appropriate to the Spanish American nations still in formation.

If what defines the relationship between *modernismo* and what has been called *posmodernismo* is their shared poetic vision based on universal analogy, a metaphoric system in which the physical world reveals corresponding spiritual realities, much of Julio Herrera y Reissig's poetic production (like that of the early Lugones) can be seen as an alternate development of *modernista* tendencies, one that anticipates the avant-garde experimentation with poetry as a verbal construct. This conception of poetic discourse is at times tragically or ironically locked within its own structures, at times defiantly antagonistic to the language of everyday reality, at times prosaic in its rejection of literary models. It remains, nevertheless, hopeful that, in its inventive commentary, the power of art can effect an impact upon its social context. Yurkievich, in the final paragraph of his insightful *Celebración del modernismo,* expresses similar conclusions:

> The extreme sublimation and stylization by Herrera y Reissig responds to a radical rejection of all utilitarianism, to a hardened disaffection with the [social] order based upon profit, to opposition to the established forms of life. His poetry acts like a distancing mediation from alienated existence; he wants to recover through distance and strangeness the transcendence that is unattainable through social interaction. It is the decrying of an absence, of a mutilation, of a missing dimension. It is subversive art: it proposes an imaginary re-creation of the factual experience. Negation of the ruling order, negation of all repressive order, Herrera y Reissig's poetry disconnects itself completely from the circumstantial and the contextual in order to preserve a freedom that can only exist in the realm of the aesthetic. Aesthetic values, even though unrealizable, imply a rebuke of dominant values. . . . Poetry . . . takes refuge in an illusionist integrity, in a universe of exalted fiction in order to oppose it to the impoverishing violence of the practicable world. (97–98)[13]

While the creative leap taken by Herrera y Reissig should not be underestimated, neither should its footing in the works of his *modernista* predecessors be overlooked. As I have shown throughout this study, from the very beginning *modernista* art has offered itself—even in its most "fantastical," "exotic," or "decadent" pieces—as a response to its immediate social circumstances. Herrera y Reissig absorbed the profound lessons of the *modernista* legacy and took the movement a step further. The early *modernista* desire to create a poetry of spiritual and political transcendence based on "the poetic" finds a fertile development in the verbal options opened by Herrera y Reissig. He builds upon what previous poets had made possible in Spanish—recourse to verbal correspondences and synaesthesia, the incorporation of vocabulary and images from high culture, and the descriptive renditions of wondrous landscapes. Yet he pushes these elements out of proportion and beyond their bounds. Accordingly, what comes to predominate in his poetry is tension and rupture, a sense of being off balance that reflects an underlying fear that the goals that had been set up by his predecessors were no longer attainable. As doubt erodes confidence, parody, irony, and discord replace harmony and unity. Herrera y Reissig's lifework, never completely in one camp or the other, must therefore be considered as a whole, with different poems of diverse natures commenting upon and critiquing each other and those that came before.

While other *modernistas* struggle to reveal, through analogy, the divine intelligence that imbues the universe with order, Herrera y Reissig expands verbal possibilities in such a way that the metaphoric connection between things is less direct. For Herrera y Reissig, metaphors are not consistently based on external realities but rather appear as projections of his sophisticated readings or of his—not always stable—psychic states (hence the repeated allusion to *esplín* [spleen], *jaquecas* [migraines], *neurosis* [neurosis], *neurastenias* [neurasthenias], and *lo espectral* [the phantasmagorical]). As metaphoric links become less natural, they appear more elusive, surprising, arbitrary, imaginary, ironic, and subtly or violently critical. The effect is often shocking and deliberately unaesthetic. Whether this poetic vision stems from a tragic awareness, starting at age five, of a congenital heart lesion that would doom him to a short life, or from a rebellious anarchistic, antibourgeois dandyism that would keep him on the margins of society (looking out at the world from within his *Cenáculo* [Guest Chamber], later called the *Torre de los Panoramas* [Tower of Panoramas]), or from an eclectic, judgmental, and willful character that sought to push his native Montevideo out of its provincial and unimaginative conservatism, Herrera y Reissig puts *modernista* tendencies under a magnifying glass, perfecting, exaggerating, and distorting what he passionately embraced.

It is difficult to trace a clear chronological trajectory of Herrera y Reissig's poetic development because of the way he chose to publish his work.

He organized only one collection of poems, *Los peregrinos de piedra* [*The Stone Pilgrims*] (1909), which appeared posthumously the year of his death. It is an anthological compilation of his entire oeuvre, but he excluded from it pieces that are as important and well executed as those he included. For example, he left out a third of the poems from *Los éxtasis de la montaña* and more than half of those from *Los parques abandonados* [*The Abandoned Parks*]. To the extent possible, scholars have attempted to date individual poems and group them within their original collections, but the dating is often imprecise because Herrera y Reissig, it is said, worked on a number of collections at the same time and revised them over time. What can be safely stated, however, is that most of his poetry belongs to the ten-year period between 1900 and 1910. His early poems belong to the first three or four years of this decade. Those written before 1900 are generally considered his weaker, less original pieces.[14]

As might be expected, the poems from 1900, especially those from *Las pascuas del tiempo* [*The Festivals of Time*], strongly reflect the influence of other *modernista* poets, particularly Darío, Lugones, and, most notably, Casal, whose decadent dandyism finds parallels in Herrera y Reissig's lifelong attraction to images of disease and psychoses. These poems reveal the same preoccupation with elegance, opulence, and European culture that molded much of early *modernismo.* The famous "Fiesta popular de ultratumba" ["Popular Party from Beyond the Grave"], with its wide-ranging exploration of styles and modes of discourse, recalls *Prosas profanas,* especially its first three poems. Though many of Herrera y Reissig's pieces present a cultural smorgasbord similar to Darío's "Divagación," the Uruguayan's ironic perspective is already present in his inversion of the generally solemn deference given by fellow *modernistas* to imported high culture. "Fiesta popular de ultratumba" exemplifies Herrera y Reissig's humorously mocking tone. In stanza 17 he writes:

> Un estoico de veinte años, atacado por el asma,
> se hallaba lejos de todos. "Denle pronto este jarabe,"
> dijo Hipócrates, muy serio. Byron murmuró, muy grave:
> "aplicadle una mujer en forma de cataplasma." (142)

[A stoic young man of twenty, attacked by asthma, / found himself far from everyone. "Quickly give him this syrup," / Hippocrates said, very serious. Byron, very grave, murmured: / "apply to him a woman in form of a cataplasm."]

Here Herrera y Reissig wittily wages a savage attack on numerous commonplaces: inherited wisdom in the form of high culture, literary role models, the function of sexuality in the "health" of the universe, and the "seriousness" of the entire artistic enterprise. In the final stanza he is even stronger in this indictment of *modernista* assumptions as delirium appears at the party and seems to be given the final word.

> Todos soltaron la risa. (Grita un paje: está Morfeo).
> Todos callan, de repente . . . todos se quedan dormidos.
> Se oyen profundos ronquidos.
> (Entra en cuclillas un loco que se llama Devaneo). (142)

[Everyone lets loose with a laugh. (A page shouts: Morpheus is here). / Everyone gets quiet, suddenly . . . everyone is asleep.
Deep snores are heard. / (Squatting down a crazy man whose name is Delirium enters).]

With the final stanza the poet appears to suggest that the cultural excesses of the party—even when one knows not to take them too seriously—may lead to a destructive frenzy that is counterproductive and may bring about an intellectual surfeit not unlike what happens in some short stories by Borges.

Though the other poems of *Las pascuas del tiempo* are similarly filled with historical, mythological, and literary names and references, his next collection, *Los parques abandonados,* follows a different strategy in its reformulation of *modernista* tendencies. *Los parques abandonados* consists of poems written between 1900 and 1908. Twenty-two were published in *Los peregrinos de piedra* under the newly coined heading of "Eufocordias"; fifty-seven were left out. As Allen Phillips has noted, Herrera y Reissig reveals a complicity among the elements of the creation in these gentle and touching love sonnets ("Cuatro poetas" 438). In "La estrella del destino" ["Destiny's Star"] (39), the responsive character of the universe is captured in the image of a star that speaks to a bereaved lover. The star's powers are linked to the deceased, in whose eyes the light of the heavenly body had found a complement. In "La sombra dolorosa" ["The Painful Shadow"] (41), nature shares with the poet the pain of separation, affirming a transcendental unity which is then interrupted by a blaring train that simultaneously asserts and bemoans the dislocations of modern life. In "El abrazo pitagórico" ["The Pythagorean Embrace"] (80), the universal and eternal harmony alluded to in the title is found in the union of the two lovers, their love, their song, and their pleasure. The poem is not, however, without a Herrerian touch of absurdity in the use of intrusive scientific terms (the rhyme of "sulfuro" and "bromuro"). Throughout these poems and others in the collection, Herrera y Reissig underscores the fragility of the vision he pursues, disclosing that behind the bright promise of accord lurks a specter of pain and rupture, which is often the product of modernity.

The poems of *Los éxtasis de la montaña,* called "Eglogánimas" in *Los peregrinos de piedra,* date from 1904 to 1910 and are much more uniformly optimistic. These miniature masterpieces are pastoral in nature, eclogues full of peace and tranquility. The poems reveal a delight in the innocent and ingenuous nature of rural life and express an intense pantheism in which the cosmic order manifests itself in apparently insignificant details of everyday existence. The first "Eglogánima" of *Los peregrinos de piedra,* "El despertar" ["The Awakening"], is a magnificent example of these characteristics. It is an expertly executed alexandrine sonnet whose joyful faith in the perfection of the untouched landscape rivals Martí's *Versos sencillos* and Darío's pantheistic poems of "Las ánforas de Epicuro," especially "La espiga" ["The Ear of Wheat"].[15]

Alisia y Cloris abren de par en par la puerta
y torpes, con el dorso de la mano haragana,
restréganse los húmedos ojos de lumbre incierta,
por donde huyen los últimos sueños de la mañana . . .
La inocencia del día se lava en la fontana,
el arado en el surco vagaroso despierta
y en torno de la casa rectoral, la sotana
del cura se pasea gravemente en la huerta . . .
Todo suspira y ríe. La placidez remota
de la montaña sueña celestiales rutinas.
El esquilón repite siempre su misma nota
de grillo de las cándidas églogas matutinas.
Y hacia la aurora sesgan agudas golondrinas
como flechas perdidas de la noche en derrota. (9)

[Alisia and Cloris open the door wide / and clumsily, with the back of their idle hand, / they scrub their moist eyes of uncertain light, / through which the last dreams of the morning flee. . . .

The innocence of the day bathes in the fountain, / the plow in the roving furrow awakens / and around the rectory, the priest's / cassock strolls seriously in the orchard. . . .

Everything sighs and laughs. The distant calm / of the mountain dreams celestial routines. / The cattle bell always repeats its same cricket-like

note of the guileless morning eclogues. / And toward the dawn the sharp swallows cut a bias / like lost arrows belonging to the defeated night.]

This timeless poem deliberately steps out of history to present an eternal image void of worldly confrontations. Like a PreRaphaelite painting, the pristine world portrayed reveals what is impossible to perceive in the urban clutter and disorder of Montevideo and the other Spanish American capitals at the turn of the century. In the countryside the poet is free to perceive celestial patterns and to apprehend, in the flight of swallows, the victory of light over dark, day over night, hope over fear. The sonnet evokes a spiritualized conception of all that has been hidden or destroyed in an urban, industrial world.

On reflection, however, the idealized nature, that is, the literariness of these images, gives the scene an unreal quality. A tension emerges between natural grace and artistic re-creation, which in the end hints at the distance between the imagined vision and the reality at hand. The landscape is transmuted into a verbal construct that, to a certain extent, coincides with Baroque sensibilities and reflects the influence of Góngora.[16] The presence of the controlling artist destroys the illusion of an unmediated apprehension of reality. This disjunction is foregrounded even further in other sonnets of *Los éxtasis de la montaña* in which the fantasy atmosphere is interrupted by the absurdity of life or by a shocking image. "La iglesia" ["The Church"] (13) ends with a flood of chickens, "El cura" ["The Priest"] (13–14) concludes with the priest's piety described as a cow's lick ("Y su piedad humilde lame como una vaca"), and "Dominus vobiscum" (64) disrupts the timelessness of the countryside with the appearance of a "zootécnico," a professor of worms.

While most *modernistas* believed that poets could translate the book of nature flawlessly by becoming attuned to the rhythms of the universe and by opening their souls completely and sincerely to the perfection of existence, Herrera y Reissig, through poems like those of *Los éxtasis de la montaña,* hints at the unavoidable presence of social and artistic obstacles that hinder the pursuit of that goal. The imagined medieval purity of the provinces, which is associated with the church and rural religiosity in the poems discussed, is made to confront the intrusive realities of life and the inevitable distortions of the ideal. Herrera y Reissig states his disillusionment with *modernista* aspirations in a more emphatic manner in pieces like those in *La torre de las esfinges* [*The Tower of the Sphinxes*] (27–38), written in 1909 and accurately if self-mockingly subtitled "Psicologación morbo-panteísta" ["Morbus-Pantheistic Psychologation"]. Within this series of poems there are enduring, if erratic, impressions of a pantheistic universe and allusions to numerous esoteric and Eastern religions. These references, however, appear to float unanchored in a nightmarish vision of rupture and death. In the seventh stanza of the first section, "Vesperas" (27–30), Herrera y Reissig comments upon his poetic juggling act with regard to the "Great Everything":

> Fuegos fatuos de exorcismo
> ilustran mi doble vista,
> como una malabarista
> rutilación de exorcismo . . .
> Lo Subconsciente del mismo
> Gran Todo me escalofría
> en la multitud sombría
> de la gran tiniebla afónica
> fermenta una cosmogónica
> trompeta de profecía. (28–29)

[Exorcism's swamp fires / explain my doble vision, / like a juggling / brilliance of exorcism. . . . / The Subconscious of the very same / Great Everything gives me chills / and in the somber multitude / of the great aphonic darkness / a cosmogonic trumpet / of prophesy ferments.]

Herrera y Reissig is juggling the *modernista* view that the universe is imbued with life and music with his perception that it is now silent and, therefore, incapable of communicating meaning. He wavers between the faith that the poet is a prophet, on one hand, and the fear that prophesy is no longer possible and that the poet can only make copies of images already created in art and literature, on the other. The latter position, which resonates with postmodern overtones, appears in the confession that he finds things to be facsimiles of his hallucinations: "Las cosas se hacen facsímiles / de mis alucinaciones / y son como asociaciones / simbólicas de facsímiles . . ." ["Things become facsimiles / of my hallucinations / and they are like symbolic / associations of facsimiles . . ."] (31). With this statement Herrera y Reissig reveals a coincidence between the course of his own work and the trajectory of the *modernista* movement (see chapter 1). Unable to trust his ability to perceive and render in art the ordered beauty of the universe, he begins to question the correspondence between poetic language and the realm that transcends worldly appearances. He fears that his, perhaps all, perceptions are "always already"—in a Derridean leap—prestructured; he suspects that we live in a world of ready-made perceptions. Herrera y Reissig remained a *modernista,* in love with demanding poetic structures—the sonnet and the *décima*—fond of rich and exotic landscapes and tonalities, and a disciple of the symbolist faith in evocation and suggestion. Yet there can be little wonder that he has been recognized by many later writers—César Vallejo, Vicente Huidobro, Pablo Neruda, Federico García Lorca, and Vicente Aleixandre, to mention only the most prominent—as a kindred spirit and teacher.

Whether she intended to or not, Delmira Agustini, like Lugones and Herrera y Reissig, explores the limits of the *modernista* conceptions of existence and poetry. Her life and career can be summarized with a few short sentences that belie the power and complexity of her legacy. She began to publish poetry in small journals at the age of sixteen. By 1907 she had published her first volume of verse, *El libro blanco* [*The White Book*]. In 1910 her second collection of poetry, *Cantos de la mañana* [*Songs of the Morning*], appeared, and in 1913 she published *Los cálices vacíos* [*The Empty Chalices*] with an opening poem by Rubén Darío. In this 1913 publication she announced her next book, *Los astros del abismo* [*The Stars of the Abyss*], which would appear posthumously in 1924 with the title *El rosario de Eros* [*The Rosary of Eros*]. Agustini was killed in 1914 by her ex-husband, whom she had taken as a lover.

Because she is so much younger than the other *modernista* poets, she is often classified—along with Alfonsina Storni, Juana de Ibarbourou, and Gabriela Mistral—as a *posmodernista.*[17] It is true that she, together with her compatriot and contemporary María Eugenia Vaz Ferreira, shared with these later women poets innovative perspectives on art, love, feminine sexuality, and the role of women that affected her imagery and tone and that distanced her from the earlier male *modernistas.* Yet it is equally significant that she maintained during her short and turbulent life a strong affinity for the principal elements of *modernista* verse. Her vocabulary, like that of other *modernista* poets, is at different moments sensuous, ornate, evocative, and exotic. Her descriptions are textured by subtly nuanced adjectives and by references to precious stones, flowers, animals, and opulent objects. Her images affirm their *modernista* roots in their detail, power, and sensuality but reject established patterns of (male) perception; they are innovative and dramatic, based on shocking connections that assert the individual and idiosyncratic nature of her experience. Her challenge to and struggle against imposing and inflexible structures—social and poetic—appear most often and most explicitly in her pieces on love and art and in her repeated allusions to sadness, suffering, and death.

It has been said that love was Agustini's great theme, yet the love that she portrays is neither simple nor idealized. Her concept of love embraces both pain and pleasure, surrender and rejection, power and impotence. Whether this dark and tortured vision is grounded in decadent art, sadomasochistic impulses, or a belief in the voluptuousness of death (as proposed by Doris T. Stephens), it suggests, in an uncannily prophetic manner, that love entails suffering and destruction and that love ultimately demands a breaking out and away. What is questioned and challenged in the process are dominant values, normative expectations, traditional male (and *modernista*) perspectives on sexuality, authority, and transcendence. This impulse to wage war on the social and poetic status quo pervades Agustini's views, not only on love but also on art, and moved her in the direction of the avant-garde.

While Agustini shares the *modernista* aspiration to an art that discloses the underlying meaning of the universe and that is redemptive, she affirms, at the same time, the suffering and despair that are inherent to the artistic process as the poet confronts time-honored models, seeks to express the ineffable, and aspires to exceed human, poetic, and stereotypical limitations. As Sylvia Molloy has brilliantly shown, in Agustini's "Nocturno" ["Nocturne"] *modernista* art—vocabulary, setting, and tone—is ritualized, only to be exploded in the surprising, touching, and frightful inversion of the image of the swan. Agustini identifies herself as the swan that stains the lakes with its blood as it tries to take flight.

Engarzado en la noche el lago de tu alma,
Diríase una tela de cristal y de calma
Tramada por las grandes arañas del desvelo.
Nata de agua lustral en vaso de alabastros;
Espejo de pureza que abrillantas los astros
Y reflejas la sima de la Vida en un cielo! . . .

Yo soy el cisne errante de los sangrientos rastros,
Voy manchando los lagos y remontando el vuelo. (254)

[The lake of your soul is encased in the night, / one would say a cloth of crystal and of calm / woven by the great spiders of watchfulness.
The best of purifying water in an alabaster vessel; / Mirror of purity you make the stars shine / and you reflect the abyss of Life in a heaven! . . .
I am the errant swan of the bloody trails, / I go on staining the lakes and taking flight.]

With her blood, Agustini writes upon the soul of the other, who, for Molloy, is Darío. She inscribes herself within his framework in a violent reaction against the stagnant *modernista* vision of life, love, and sexuality. If the *modernista* poem makes the stars shine and turns the abyss of life into a heaven, Agustini is not content to remain passively adrift upon its surface. She struggles fiercely to create a space for herself in a poetic universe in which, as shown in chapter 4, woman is repeatedly silenced in favor of the male voice. She feels compelled to expose the wished-for vision of the *modernistas* and to reveal an underlining reality that is less than perfect. The dangers of such an ideological break are many. As a result she suffers in this poem and in others. As bleeding swan, wild beast, vampire, or angel who has lost her wings, she imagines herself paying the price for confronting poetic and social norms.

Significantly, the marker for her rebellious stance is also the marker for her gender. The blood of her suffering is also the blood of her womanhood—monthly cycle, loss of virginity, childbirth. She therefore conflates her punishment with her sex and her femaleness with her struggle. Her writing stains the smooth surface of patriarchal discourse which calmly reflects its own version of the universe. Her creativity leaves a trail of both pain and joy as she attempts to escape customary limitations.

Agustini's poetry thus offers a revealing commentary on the movement whose spirit she embraced while rejecting those features that had become predictable and routinized. Unlike writers like Chocano, who patterned their poetry on the predominant rhetorical qualities that converted *modernismo* into a popular, mainstream force in the cultural life of Spanish America, Agustini discovers within *modernista* discourse the critical power to subvert the images and concepts that had become comfortable and conventional. The impetus that she finds is fueled by the same explosive mix of social change and poetic dissatisfaction that stimulated the creation of *modernismo* itself. Pressures on the hierarchical relations within society alter assumptions about the role of women. Agustini responds to these shifting patterns by presenting a dynamic critique of the dominant male perspective. Her assault on the views of the old guard was to attract followers who further advanced the orientation that she articulates. Storni, Ibarbourou, Mistral, and later Castellanos all acknowledge Agustini's influence (see Beth Miller 19–20). These women writers turn away from *modernista* formulas that had become exhausted from overuse. They, nonetheless, like Agustini, remain true to *modernismo*'s epistemological and political goals.

The demise of *modernismo,* however, does not simply result from these continuing shifts and adjustments within the *modernista* camp. Around 1920 the *modernista* framework falls under attack from outside forces. While some writers, such as the surrealists and *creacionistas,* hold fast to the faith in a comprehensible universe, *modernismo*'s optimistic worldview, formulated under the syncretic influence of the occult sciences and based upon ancient beliefs in the harmony of the universe, no longer appears viable. More significantly, the inequities glimpsed by Martí take center stage as the Great Powers go to war and the repercussions of the global economy bring fiercely destructive patterns of exploitation to Spanish American life. Literature takes on an aggressive posture that is, on one hand, incompatible with *modernismo*'s search for harmonious beauty but is, on the other, completely consonant with *modernismo*'s hope to be a collaborative force in Spanish America's creation of a modern vision of itself.

SEVEN *Modernismo*'s Lasting Impact

THE basic argument that I have articulated in this book has been that *modernismo* represents Spanish America's first full-fledged intellectual response and challenge to modernity. The resulting dialogue with predominant modern values and tendencies was, on the surface, complex and diverse, because it was shaped by a multiplicity of conditions—including personal, political, social, and philosophic factors. For many years, the outwardly disparate nature of the movement was the focus of critical thought about *modernista* texts and authors, obscuring a wide-reaching vision of common concerns and issues that grounded the movement and converted it into a foundational force. Yet by emphasizing *modernismo*'s probing and profound reaction to Spanish America's troubled entrance into the modern era, I have sought to demonstrate the movement's unity as well as its crucial role in granting literature the authority to address and influence the forces of modernity and to map the course of Spanish American development.

Modernismo, like the modern European movements that preceded it, claimed for literature an adversarial function. It sought to provide a vision of ultimate truths that operated on two levels. It strove to offer spiritual solace and transcendental significance by reaching out for alternatives with which to replace those religious beliefs undermined by critical reason and positivistic thought. It also aspired to counter the superficial and vacuous concerns promoted by the materialistic and pragmatic values of modern life. The *modernistas* sought to achieve this vision by creating a revitalized language that turned its back on inflexible patterns of perception and expression at the same time that it embraced beauty from all eras and all regions. This language was to resonate with the harmony of existence and to achieve an accord with the primordial, prelapsarian perfection of the universe, thereby entitling the authors to reorient the moral compass of their countries.[1]

As hack imitators hollowed out the verbal structures that the *modernistas* had constructed, and as art and artifice became progressively more tightly linked to artificiality and insincerity, readers and critics alike tended to overlook *modernismo*'s confrontation with modernity and the political and epistemological response that it provoked. They failed to understand that *modernistas* believed that the truth of poetry was revealed in—not hidden behind—ornamentation. They focused on the well-crafted form of *modernista* poetry and its attention to musicality and disregarded the *modernistas*' serious encounter with the social and political realities of the day. Here I have set out to reclaim the power of the visionary stance taken by these creative intellectuals through systematic consideration of the development of the movement together with detailed analyses of works that typified and altered its course.

My analysis of this trajectory clarifies how *modernista* discourse matured, acquired a sense of self-worth, and became simpler, more direct. These stylistic shifts, which were often seen as departures from *modernismo,* are part of the evolution of the movement. At the same time, however, some writers began to question the ability of poetic discourse to fill the voids left by the disruptions of modern life. As doubt moved from periphery to center, irony broke through the carefully crafted surfaces of the poetry. This development began with Lugones and Herrera y Reissig and emphatically continued into the avant-garde.

The *modernista* rejection of a rigid sense of "realism" in favor of a more imaginative, spiritual, even oneiric, orientation further facilitated the transition to the avant-garde. In their search for knowledge, *modernista* writers explored numerous religious, mythic, and symbolic systems. The interplay of these various referential codes also opened new linguistic possibilities. No longer fixed, signs and symbols gained greater and greater independence, increasing the free play between signified and signifier. The move from *modernismo* to the avant-garde to the deconstructionist discourse of today's postmodernism thus appears to be more of a gentle slide than a series of cataclysmic realignments.

Even more essential to these developments, however, is *modernismo's* faith in the transformative capacity of art, that is, its ability to see beyond the stultification of everyday existence and entrenched worldviews and to perceive truths that can act as correctives to human foibles and failed social systems. For Octavio Paz, this antagonistic perspective lies at the heart of modern art in general. For Paz, "modern art is not only the offspring of the age of criticism, it is also its own critic" (*Children* 3).[2] In other words, modern art positions itself both against modern life and against the art that immediately precedes it. The "modern tradition" is therefore an "interruption of continuity" or a "tradition against itself" ("la tradición de la ruptura"). The constant renovation of poetic form associated with *modernismo* and its metamorphosis into *posmodernismo* or into the avant-garde are also features derivative of its modernity. All these movements, despite their formal differences and generational tensions, maintain a fundamental and overriding concern that intellectual production provide a "critical" counterbalance to the dominant trends of world capitalism and the concomitant alterations in daily existence.

Some critical works and histories have recognized the impact of *modernismo* on later movements. Yet, by emphasizing aspects of *modernismo* that have remained in the shadows, I have attempted to show that its legacy is actually broader than the one usually observed. From the perspective outlined in the previous six chapters, the aggressively rebellious stand taken by a poet like César Vallejo can now be recognized as elaborating upon the social commentary and poetic experimentation begun among the *modernistas.*[3] The prophetic stance adopted by Pablo Neruda and his desire to rewrite "traditional" versions of Spanish American history also reflect the influence of the visionary role assumed by the *modernistas.*[4] Even Paz's fusion of historical reconstruction with epistemological concerns about falsehoods and masks presented in *El laberinto de la soledad* [*The Labyrinth of Solitude*] suggests the lasting impact of the *modernista* project.

This creative debt has been acknowledged both explicitly and implicitly by numerous writers. I have chosen to offer a mere sampling in the following pages in order to highlight the presence of *modernismo* in some of the most important texts of the second half of the twentieth century. The further elucidation of this relationship in these and in the many other works in which it appears is left, more appropriately, as the focus for another study.

One of the most direct allusions to the foundational contribution of *modernismo* occurs in *Rayuela* [*Hopscotch*] by Julio Cortázar. A classic of self-referentiality, overwhelmingly concerned with defining itself, the nature of literature, and the role of the author, *Rayuela* provides a novelist's view of recent literary history.[5] By mentioning González Martínez's "Tuércele el cuello al cisne . . ." at a crucial moment in the novel, Cortázar reveals his sense of filiation to *modernismo* and its aspirations (*Rayuela* 238–253; *Hopscotch* 213–227).[6]

In the last chapter of the first part of *Rayuela,* Horacio Oliveira, the protagonist of the novel, walks through the streets of Paris reevaluating his experiences there. This vibrant metropolis was and continues to be an intellectual mecca for Spanish Americans, but in the novel it evokes the type of magical, empowering cultural center that it came to represent for the *modernistas.* In the byways of that great city, Oliveira reflects upon the search that has dominated both his life and the previous pages of the novel. He is about to concede that his search for the "kibbutz of desire" has been a failure and that "perhaps victory was to be found in that very fact" (*Hopscotch* 214).[7] He has come to realize that the kibbutz is a fortress that can only be taken "with the aid of arms contrived in fantasy, not with the soul of the West or with the spirit" worn away by lies (*Rayuela* 215).[8]

This general statement reasserts the *modernista* belief that Western science and its empiricist epistemology have ignored the most important aspects of existence. The declaration also repeats the *modernista* despair over the weakness of "the spirit" that has been contaminated by the trivialities of daily life. Poetic imagination, fantasy, and artistic vision appear as the sole source of hope, the only way to escape the distortive and constraining social and cognitive patterns that have closed modern individuals off from profound wisdom. At this critical juncture, Oliveira enters the underworld of the *clochards,* the homeless, and threads throughout his conversation with the foul-smelling Emmanuèle references to González Martínez's most famous poem. Through this recourse to "Tuércele el cuello al cisne . . . ," Cortázar acknowledges the *modernista* project as a direct antecedent not only to Oliveira's search but also to his own most profound intellectual struggles.

These struggles include an explicit political dimension which Cortázar felt compelled to address directly at various points throughout his career.[9] In the novel, issues of national identity and national destiny are brought to the fore a few pages later, with Oliveira's return to Argentina. His return in the second section of the novel is preceded by an epigraph by Apollinaire. The quote resonates with meaning: "Il faut voyager loin en aimant sa maison." This assertion recalls the countless Spanish American intellectuals who, like Oliveira, Cortázar, and many *modernistas,* traveled abroad (physically or imaginatively) in search of knowledge to better understand who they are and how best to deal with the numerous dilemmas of modernization confronting their homelands.

In a similar fashion, Gabriel García Márquez gives a literary nod to *modernismo* and its encounters with the predominant materialist ideology at the beginning of *Cien años de soledad* [*One Hundred Years of Solitude*]. The first pages of this masterpiece present Macondo and the Buendía family as defined by a fundamental tension between scientific knowledge and spiritual understanding, a tension that is generated by the misguided appropriation of foreign inventions. Melquíades brings to the Edenic Macondo an array of items that jolt it out of its timelessness and into an undigested modernity. In turn, the modern goals set up by José Arcadio Buendía distract him from his original obligations of nation building, leaving community formation in the hands of the intuitive Ursula. In reaction to José Arcadio's unthinking, selfish, and greedy utilization of the imported treasures, Melquíades repeats *modernista* injunctions against the insensitivity of the ruling classes and the limitations of positivist perspectives. "'Things have a life of their own,' the gypsy proclaimed with a harsh accent. 'It's simply a matter of waking up their souls'" (*One Hundred Years* 11).[10] The dual *modernista* concerns of epistemology and politics once again intertwine as García Márquez expresses his indebtedness to *modernismo* and its inaugural battles with the forces of modernity. *Cien años de soledad* can, consequently, be understood as a reassessment of Spanish American history from the perspective of *modernismo,* specifically, from the perspective of the *modernista* rejection of those decisions by the dominant classes that locked Spanish America within what is portrayed as an egomaniacal downward spiral toward an apocalyptic end.

While the importance of García Márquez's intertextual tribute to *modernismo* in *Cien años de soledad* may appear diluted by similar—though less strategically placed—allusions to the *crónicas* and even to the *novela de la tierra,* it is emphatically reasserted eight years later with the publication of *El otoño del patriarca* [*The Autumn of the Patriarch*] in 1975. The name Rubén Darío virtually brackets the novel, appearing four pages from the beginning and from the end of the text.[11] With these references, which are supported by other allusions throughout the novel, García Márquez once again suggests that the origins of the modern Spanish American state as well as of the prototypical Spanish American dictator are to be found in the region's initial encounter with modernity during the second half of the nineteenth century. He emphasizes, through hyperbolic example, how political repression is an outcome of foreign economic encroachments similar to those that began to take place during the *modernista* period. More importantly, however, the figure of Rubén Darío is presented as the source of that other, "mystical" voice, for the most part unheard in the cacophony of self-serving political lies.

Darío's poetic voice is heard one magical evening in chapter 5 during an evening of poetry at the National Theater.

> . . . during the two hours of the recital we bore the certainty that he was there, we felt the invisible presence that watched over our destiny so that it would not be altered by the disorder of poetry, he regulated love, he decided the intensity and term of death in a corner of the box in the shadows from where unseen he watched the heavy minotaur whose voice of marine lightning lifted him out of his place and instant and left him floating without his permission in the golden thunder of the trim trumpets of the triumphal arches of Marses and Minervas of a glory that was not his general sir. . . . (*Autumn* 180–181)[12]

The fictionalized reading of Darío's "Marcha triunfal" leaves the dictator feeling overwhelmed by the power of poetry. He

> felt poor and tiny in the seismic thunder of the applause that he approved in the shadows thinking mother of mine Bendición Alvarado this really is a parade, not the shitty things these people organize for me, feeling diminished and alone, oppressed by the heavy heat and the mosquitoes and the columns of cheap gold paint and the faded plush of the box of honor, God damn it, how is it possible for this Indian to write something so beautiful with the same hand that he wipes his ass with, he said to himself, so excited by the revelation of written beauty. . . . (*Autumn* 181)[13]

Though the potential impact of this sudden burst of poetic inspiration is blown away almost immediately by the dynamite that destroys the dictator's small family, García Márquez holds out the possibility proposed earlier by Rodó and other *modernistas,* namely, that beauty might prove a powerful teacher and that the lessons learned might enhance the moral stature of those who hold power. He also recognizes, however, that in the current climate, poetry teeters between total oblivion and easy recognizability (that is, between "the forgotten poet" of the beginning of the novel and "the famous poet Rubén Darío" of the end). In this way, García Márquez acknowledges in his own work and life a continuity in the cultural battles of modernity that began with *modernismo.*[14]

Alejo Carpentier's *El recurso del método* [*Reasons of State*] is his clearest homage to *modernismo.* As brilliantly shown by Roberto González Echevarría, the novel acknowledges the movement as Spanish America's first and grounding confrontation with modern life ("Modernidad"). Yet the continuing legacy of the *modernista* project is also evident in *El siglo de las luces* [*Explosion in a Cathedral*] and in *El acoso* [*The Chase*]. The former plays out the struggle that underlies many *modernista* works, that is, the tension between defunct Enlightenment aspirations (highlighted by the reference in the Spanish title to "the century of lights") and spiritual alternatives that are offered to take their place. As in *modernismo,* occultist sects and symbols imbue these alternatives, as well as the literary text itself, with transcendental value. In *El acoso,* this compensatory search for the lost paradise of beauty and truth is simultaneously spiritual and political and follows the romantic paradigms that structured the *modernista* worldview.[15]

Though *La campaña* [*The Campaign*] by Carlos Fuentes focuses on the period preceding *modernismo,* it also takes up the *modernista* battle for the hidden soul of Spanish America against the forces of the Enlightenment, critical reason, and foreign thought. The mystical search for Eldorado, the ideological wars for control of the continent, and Baltasar Bustos's confrontation with the barbarous and instinctual "other" in the guise of the gaucho all come together in an attempt to recapture the essence of Spanish America and to redefine the course of modernity. In an article appropriately entitled "La literatura es revolucionaria y política en un sentido profundo" ["Literature is Revolutionary and Political in a Profound Sense"], Fuentes spells out this dual epistemological and political vision in terms that recall the romantic, symbolist, and *modernista* roots of contemporary Spanish American literature. He writes: ". . . literature not only sustains a given historical experience, it not only continues a tradition, but, through moral risk-taking and formal experimentation and verbal humor, it breaks the conservative horizon of readers and contributes to freeing all of us from the chains of an ancient way of seeing, from a sterile womb, from a musty and doctrinaire prejudice" (*La literatura* 14).[16]

More radical but no less aware of its intellectual affiliation with *modernismo* is the work of Severo Sarduy. In 1967, the centenary of Darío's birth and the year in which Sarduy published his groundbreaking *De donde son los cantantes* [*From Cuba with a Song*], he wrote a discerning if somewhat self-reflective article on the Nicaraguan poet. In it he stated: "In the perception that Darío has of the world as a signifying code there is an intermediary. That intermediary is *always* of a cultural order, that is to say, that Darío introduces into literature this fundamental dimension . . . : the poem is situated in a sphere that is completely cultural, in what the structuralists call 'the code of paper.' This intermediary, always plastic in [Darío], is frequently also of a literary order, and when I say literary, I mean Verlaine" (Sarduy, Segovia, and Rodríguez Monegal 36).[17]

As is clear from this statement, Sarduy understood that Darío, by passing his perceptions through an artistic filter, actually "dresses" his concerns in a style that he associates with the prestige and elegance of the Old World and ancient cultures. Darío's desire to see his poetic vision "masquerading" (as in "Era un aire suave . . .") at the same symbolic feast as that of his great idols is an antecedent to Sarduy's creative (and iconoclastic) transvestites—perhaps most notably Auxilio and Socorro from *De donde son los cantantes.* Darío's female other—muse, poetry, poetic discourse—starts out changing her traditional and conservative Spanish attire for the less constraining fashions of French couture so that she may be free to do new things and be perceived in a new way. Sarduy's transvestites are not as removed from this position as one might first suspect. They also call attention to their evolving cultural context. Like an updated version of the *modernista* vision, the world that they inhabit reflects the tensions between the old and the new, between the native and the foreign, between traditional images of national identity and the impact of the mass media and world capitalism. The transvestite struggles to locate identity in and through adornment. Moreover, for Sarduy, versed in all aspects of structuralist and poststructuralist theory, fashion operates, at least in part, as a function of the global marketplace, in which local concerns, personal identity, and non-commercial values continue to lose ground in the multinational battle begun with the arrival of modernity.

These connections are seldom, if ever, drawn because Sarduy's work, quite correctly, appears light years removed from the poetry of Darío and that of other *modernistas.* This distance, paradoxically, underscores both the continuities and the breaks between *modernismo* and its literary descendants, including postmodernism. As I have indicated throughout this study, the desire to respond to the predominant values, assumptions, and discourse of modern society informs the *modernista* project as well as the literary movements that follow in its wake. Yet, by the second half of the twentieth century, faith in the power of art to offer a viable alternative to the distortive constraints of the dominant and increasingly global culture began to be eroded by the awareness that even artists cannot completely extricate themselves from these realities. Instead of providing writers with stability and security, language and knowledge began to be seen as a coercive force that could not be trusted. With the resulting disbelief in "master narratives," there is a tendency to reject foundational philosophies and related totalizing beliefs. Answers are sought between positions, between the stands taken by the powerful and the weak, the privileged and the disadvantaged. This interplay, a continuing struggle against the facile assumptions imposed by others, structures Sarduy's works and shows how close and yet how far away he is from his beloved Darío. Unlike Darío, he accepts the possibility of multiple, simultaneous meanings that, instead of offering a balanced, harmonious vision, collide with one another. Like Darío, however, his efforts to establish meaning in an all-embracing cultural universe lead him to reject the confinement of the dominant falsifying orders.[18] This is *modernismo*'s legacy.

All these works, *Rayuela, Cien años de soledad, El otoño del patriarca, El recurso del método, El siglo de las luces, El acoso, La campaña,* and *De donde son los cantantes*—as well as many others—underscore the connection of contemporary literature to *modernismo.* Each in its own way reconfirms the lasting foundational nature of the *modernista* vision. Each emphasizes the need to read *modernismo* from the perspective of modernity. Each sees what some have been reluctant to acknowledge, namely, that *modernista* texts reflect a dynamic, sophisticated, and at times tragic awareness of a world in transition.

Notes

CHAPTER ONE

1. Because of the differences between Spanish American *modernismo* and the related but later Anglo-European phenomenon, modernism, I will use the term *modernismo* and its adjectival form, *modernista,* throughout this work to refer to the linguistically rich, formally innovative, and ideologically complex literary movement that began in Spanish America in the late 1870's and that lasted into the second decade of the twentieth century. The avoidance of the English equivalent makes it possible to distinguish immediately between the Spanish American and Anglo-European movements. Similarly, the uniquely Spanish American *posmodernismo,* which refers to the rather poorly defined tendencies of late *modernista* writings, will be kept separate from all associations with what is, in English, understood to be postmodernism.

2. For an introduction to the problematics of modernity and postmodernity in Latin America, see *The Postmodernism Debate in Latin America,* edited by John Beverley and José Oviedo and translated by Michael Aronna.

3. I do not wish to imply that the relationship between *modernismo* and modernity has been completely overlooked. I stand on the shoulders of previous critics who have perceived and studied different aspects of this connection. In contrast to them, however, I place it center stage in understanding the movement and its legacy.

As early as 1962, Luis Monguió explored Spanish America's increasing commercial involvement with the great industrial countries and described the *modernista* reaction against materialistic positivism as the impetus for the movement ("De la problemática"). Also, in 1965, a little-known article by Yerko Moretić discussed the penetration of imperialist forces and the co-optation of the national bourgeoisie as factors in the development of *modernismo* ("Acerca"). It was, nevertheless, critics like Paz (*Los hijos del limo* [*Children of the Mire*]), Calinescu, Rama, Pérus, and Gutiérrez Giradot, each in his or her own way, who began to reorient *modernista* studies. They emphasized that the writers of the period were not, as previously believed, aloof aesthetes uninvolved in the world around them. On the contrary, they were dealing with a number of serious contextual issues. Other scholars working along these lines were Zavala ("1898, Modernismo and the Latin American Revolution"), Yurkievich ("Rubén Darío y la modernidad"), Podesta, and Real de Azúa.

4. See the scholarly and informative article by Alfredo Roggiano. Ivan A. Schulman, in another wide-reaching article ("Modernismo/modernidad"), focuses on the development of the critical understanding of the term *modernismo,* especially as used in literary circles.

5. It is impossible to list all those that have underscored this link, for poststructuralist and Marxist perspectives have sensitized most contemporary writers to the impact of socioeconomic factors on modes of perceiving and thinking. Three of the most prominent theorists are Foucault, Jameson, and Lyotard.

6. In his broad and persuasive *Five Faces of Modernity,* Matei Calinescu points out that most American critics of twentieth-century literature make no distinction between modernism and the avant-garde. The two terms are often taken as synonymous. In Europe, however, the avant-garde tends to be regarded as the most extreme form of artistic negativism, whereas modernism is viewed as considerably less negative. Indeed, the antitraditionalism of modernism is often judged to be subtly traditional (140).

7. This feature of modernism explains why it is quite common to see modernist writings as difficult, complex, and elitist. Both Richard Poirier and M. Keith Booker point out that, by calling attention to the factitiousness of their own procedures, modernist texts strive to instruct their readers in ways that equip them to see beyond generally held, officially endorsed opinions (Booker 7). Once again the political and transcendental goals of literature interweave.

8. I am indebted to Arkady Plotnitsky, Visiting Fellow at the Robert Penn Warren Center for the Humanities at Vanderbilt University, 1994–1995, for his insightful comments regarding these three overlapping "postmodernisms": the artistic, the theoretical, and the cultural.

9. The original: "Habrá que pensar, para adivinar el camino que tomará la novela en un mundo que aún no podemos bautizar, primero en escritores como William Faulkner, Malcolm Lowry, Hermann Broch y William Golding. Todos ellos regresaron a las raíces poéticas de la literatura y a través del lenguaje y la estructura, y ya no merced a la intriga y la sicología, crearon una convención representativa de la realidad que pretende ser totalizante en cuanto inventa una segunda realidad, una realidad paralela, finalmente un espacio para *lo real,* a través de un mito

en el que se puede reconocer tanto la mitad oculta, pero no por ello menos verdadera, de la vida, como el significado y la unidad del tiempo disperso."

Fuentes returns to this point when discussing the role of myth and language in the writings of Miguel Angel Asturias and Jorge Luis Borges, in which he finds clear and direct ties with poetic discourse (*La nueva novela* 25).

10. This struggle, as mentioned earlier, has its distant roots in the Renaissance. For this reason, Fuentes, in yet another piece, confesses that, "given a choice in the matter, I have always answered that, for me, the modern world begins when Don Quixote de la Mancha, in 1605, leaves his village, goes out into the world, and discovers that the world does not resemble what he has read about it" ("Cervantes" 49). He goes on to point out that "[m]any things are changing in the world; many others are surviving. *Don Quixote* tells us just this: this is why he is so modern but also so ancient, eternal. He illustrates the rupture of a world based on analogy and thrust into differentiation" (50). It is precisely this longing to achieve an analogical vision in the hope of overcoming differentiation that creates a recognizable reverberation of kinship among Western writers starting with Cervantes. Or, as Lionel Trilling acknowledges: "All prose fiction is a variation of the theme of *Don Quixote* . . . the problem of appearance and reality" (qtd. in Fuentes 50). The problem with this vast perspective, however, is that it reduces all postmedieval literature to a type of sameness that obscures essential differences.

CHAPTER TWO

1. Darío's statement appears in "Palabras liminares" ["Liminal Words"] of *Prosas profanas* [*Profane Hymns*], one of the many poetic manifestos in which he outlines salient characteristics of the movement that he came to head. Davison, in his seminal work, labels this characteristic "voluntad de estilo" or "striving for an individual style" (28–30).

2. Two works that are representative of this approach are the studies by Erwin K. Mapes and Marie-Josèphe Faurie.

3. See, for example, the Ph.D. dissertation by Roger W. Miller.

4. One of the most comprehensive studies of sources of Darío's poetry is Arturo Marasso's *Rubén Darío y su creación poética.* Another general study is Dolores Ackel Fiore's *Rubén Darío in Search of Inspiration.*

5. Darío, for example, would write, "Amo más que la Grecia de los griegos/ la Grecia de la Francia" ["I love more than the Greece of the Greeks/ the Greece of France"] in "Divagación," *Prosas profanas,* 184.

6. Carlos J. Alonso's *The Spanish American Regional Novel: Modernity and Autochthony* suggests that the *novela de la tierra* can be read as one of Spanish America's responses to the crisis about the status of Latin American culture and modernity. He seems to be unaware, however, of the *modernista* antecedents to this response. Later in the twentieth century, works like *Cien años de soledad* [*One Hundred Years of Solitude*], by Gabriel García Márquez, and *La campaña* [*The Campaign*], by Carlos Fuen-

tes, offer perfect examples of how premodern indigenous cultures appear in opposition to, yet mixed with, modern ways of life.

7. The original: "[e]l espíritu moderno, es decir, el racionalismo, el criticismo, el liberalismo, se fundó el mismo día que se fundó la filología. *Los fundadores del espíritu moderno son los filólogos.*"

8. Alonso underscores these crises—together with the centenary celebrations around 1910—as giving rise to *la novela de la tierra.*

9. For an informative examination of this situation, see the article by Washington Delgado.

10. A recent article by Beatriz González Stephan sheds light on similar developments—"el guzmanato"—in Venezuela and their impact upon writing.

11. For more background to the socio-economic context of modernism, see Noé Jitrik and José Emilio Pacheco.

12. These strategies took many forms, including art for art's sake, eccentricity, dandyism, and decadentism. The idea of "l'art pour l'art," as conceived by the Parnassian Théophile Gautier, appealed to the poetic imaginations of the early *modernistas.* With this rallying call, they summarized the artist's renunciation of vulgar utilitarianism. Utility was associated with ugliness. Similarly, for some, like the English Pre-Raphaelites, their disgust with contemporary changes led them to believe that beauty existed only in *preindustrial* settings. The Pre-Raphaelite recourse to allegory and medievalism influenced those that sought to *épater le bourgeois.* A more radical variation was developed by the decadents as represented by Joris Karl Huysmans, whose *A Rebours* had a tremendous impact on a number of *modernistas,* most notably José Asunción Silva, del Casal, who corresponded with Huysmans in French, and Darío, who signed his "Mensajes de la tarde" ["Afternoon Messages"] with the name of the novel's protagonist, Des Esseintes.

13. I use this term because it parallels discoveries in quantum physics of antimatter, which is matter composed of elementary particles that are, in a sense, mirror images of the particles that make up ordinary matter as it is known on Earth. The "unreal" nature of antimatter parallels the antiempiricist emphasis of cultural modernity. Though its existence is "counterintuitive," antimatter serves as a basis for scientific explanations of things that could not otherwise be explained.

The term *antimodernity* also underscores the type of love-hate relationship that has existed, for nearly two hundred years, between science and the humanities, in which the language of one mimics the language of the other while decrying the inadequacies of the other. The tense interrelationship between these two supposedly diverse ways of looking at the universe began with scientist-philosophers such as Kant and has remained strong throughout the twentieth century with "fantastical" discoveries in mathematics and quantum theory. The oxymoronic label "occult sciences" reflects the uneasy balance between empirical knowledge and spiritual wisdom. Jorge Luis Borges may offer the most obvious example from the twentieth century of a writer both drawn to and severely critical of scientific method and discourse. In an interesting turn of events, some recent critical studies, such as those by Kathleen Hayles, have begun to focus on the coincidences that exist between poststructuralist literary theory and some late-twentieth-century discoveries in physics and computer science. For further explorations of the interplay between physics, mathematics, and critical theory, see the brilliant studies by Arkady Plotnitsky.

14. Articles by Arthur Natella, Yerko Moretić, and D. Fernández-Morera focus on these links.

15. For the following discussion I am particularly indebted to M. H. Abrams's "The Circuitous Journey: Pilgrims and Prodigals" in *Natural Supernaturalism: Tradition and Revolution in Romantic Literature,* which remains one of the best analyses of the way the major English and European poets of the nineteenth century differed from their eighteenth-century predecessors. This study explores how their common themes, modes of expression, and ways of feeling were causally related to the drastic political and social changes of the age. For additional details, see the first chapter of my *Rubén Darío and the Romantic Search for Unity.*

16. The critic who has most consistently and emphatically emphasized the importance of analogy to *modernista* poetics is Octavio Paz, first in "El caracol y la sirena" ["The Siren and the Seashell"] in *Cuadrivio* and later in *Los hijos del limo.*

17. The original: "La ciencia de lo oculto, que era antes perteneciente a los iniciados, a los adeptos, renace hoy con nuevas investigaciones de sabios y sociedades especiales. La ciencia oficial de los occidentales no ha podido aún aceptar ciertas manifestaciones extraordinarias—pero no fuera de lo natural en su sentido absoluto—como las demostradas por Crookes y Mme. Blavatsky. Mas esperan los fervorosos que con el perfeccionamiento sucesivo de la Humanidad llegará un tiempo en que no será ya arcana la antigua *Scientia occulta, Scientia occultati, Scientia occultans.* Llegará un día en que la Ciencia y la Religión, confundidas, hagan ascender al hombre al conocimiento de la Ciencia de la Vida."

18. Rafael Gutiérrez Giradot deals with these issues in the context of progressive secularization.

19. Like Abrams, Gwendolyn Bays, in *The Orphic Vision: Seer Poets from Novalis to Rimbaud,* and Denis Saurat, in *Literature and Occult Tradition: Studies in Philosophical Poetry,* contend that romantic and symbolist writings are permeated with cabalistic and hermetic thought. Bays adds gnosticism, manicheanism, Pythagoreanism, mesmerism, and spiritualism to the occult philosophies that influenced English, German, and French literary creation during the nineteenth century. John Senior, in *The Way Down and Out: The Occult in Symbolist Literature,* suggests that, among modern occultists, Eliphas Lévi has the single greatest influence on symbolism. See also Theodore W. Jensen's article, "*Modernista* Pythagorean Literature: The Symbolist Inspiration," in *Waiting for Pegasus* 169–179.

20. Aspects of these influences have routinely been identified as Pythagorean. It is best, however, to remember that the Pythagoreanism that had an impact upon *modernismo* was generally reinterpreted through esoteric doctrine and freely combined elements not only from historical Pythagoreanism but also from Neo-Pythagoreanism, Platonism, and Neoplatonism. Perhaps it was the emphasis of esoteric Pythagoreanism on order, harmony, and music that most immediately captured the poetic imagination of this generation of writers and led to the incorpo-

ration of a number of unorthodox tenets into what can be called the *modernista* worldview.

21. This recourse to sexuality is essentially the opposite of the domesticated erotic desire studied by Doris Sommer in *Foundational Fictions: The National Romances of Latin America.* In the "national romances" that she examines, Sommer finds that Latin American regimes "tacitly accepted the nineteenth-century pot-boilers as founding fictions that cooked up the desire for authoritative government from the apparently raw material of erotic love" (51).

22. The original: "Yo he dicho, en la misa rosa de mi juventud, mis antífonas, mis secuencias, mis profanas prosas."

23. The original: "Yo no tengo literatura 'mía'—como lo ha manifestado una magistral autoridad—, para marcar el rumbo de los demás: mi literatura es *mía* en mí; quien siga servilmente mis huellas perderá su tesoro personal y, paje o esclavo, no podrá ocultar sello o librea. Wagner, a Augusta Holmes, su discípula, dijo un día: 'Lo primero, no imitar a nadie, y sobre todo, a mí'. Gran decir."

24. Marcel Raymond in his *From Baudelaire to Surrealism* provides an excellent introduction to this subject.

25. The original: "No gusto de *moldes* nuevos ni viejos . . . Mi verso ha nacido siempre con su cuerpo y su alma, y no le he aplicado nunguna clase de ortopedia. He, sí, cantado aires antiguos; y he querido ir hacia el porvenir, siempre bajo el divino imperio de la música—música de las ideas, música del verbo."

26. The original: "Ninguno me ha salido recalentado, artificioso, recompuesto, de la mente; sino como las lágrimas salen de los ojos y la sangre sale a borbotones de la herida."

27. The original: "So pretexto de completar el ser humano, lo interrumpen. No bien nace, ya están en pie, junto a su cuna, con grandes y fuertes vendas preparadas en las manos, las filosofías, las religiones, las pasiones de los padres, los sistemas políticos. Y lo atan; y lo enfajan; y el hombre es ya, por toda su vida en la tierra, un caballo embridado . . . Se viene a la vida como cera, y el azar nos vacía en moldes prehechos. Las convenciones creadas deforman la existencia verdadera, y la verdadera vida viene a ser como corriente silenciosa que se desliza invisible bajo la vida aparente, no sentida a las veces por el mismo en quien hace su obra santa, a la manera con que el Guadiana misterioso corre luengo camino calladamente por bajo de las tierras andaluzas."

28. The original is entitled "Honores a Karl Marx" and reads as: "Karl Marx estudió los modos de asentar al mundo sobre nuevas bases, y despertó a los dormidos, y les enseñó el modo de echar a tierra los puntales rotos" (9: 388).

29. The original: "Asegurar el albedrío humano; dejar a los espíritus su seductora forma propia; no deslucir con la imposición de ajenos prejuicios las naturalezas vírgenes; ponerlas en aptitud de tomar por sí lo útil, sin ofuscarlas, ni impelerlas por una vía marcada . . . Ni la originalidad literaria cabe ni la libertad política subsiste mientras no se asegure la libertad espiritual. El primer trabajo del hombre es reconquistarse."

30. Blanca Rivera Meléndez (112–113) and Evelyn Picon Garfield and Iván

A. Schulman (79–96) discuss Martí's recourse to warrior imagery in this piece and in the contemporaneous *Ismaelillo*.

31. The original: ". . . la literatura comienza a autorizarse como un modo alternativo y privilegiado para hablar sobre la política. Opuesta a los saberes 'técnicos' y a los lenguajes 'importados' de la política oficial la literatura se postula como la única hermenéutica capaz de resolver los enigmas de la identidad latinoamericana."

32. The original: "Su economía será, por momentos, un modo de otorgar valor a materiales—palabras, posiciones, experiencias—*devaluados* por las economías utilitarias de la racionalización."

33. As already noted, Aníbal González also sees literature as privileging itself. González focuses on the relationship of literature over philology; Ramos focuses on politics. Neither one, however, recognizes the power of the ancient tradition of analogy to establish the legitimacy of this important move by *modernista* writers.

34. The original: "Hasta ahora, con excepciones muy contadas, la Academia se ha compuesto de personas adictas al trono y al altar; de hombres temerosos de Dios y de la gramática, que con igual entereza repugnan los pecados contra la ley de Dios y los pecados contra la sintaxis ortodoxa . . . La más ligera veleidad liberal, el más leve descuido en la sintaxis, un le, un lo, un soneto a Juárez, bastan para cerrar al candidato el santuario de las letras vocales y de las letras consonantes." "Lo que censuro es su intolerancia más bien política que filosófica." See also "La crítica literaria en México: Nuestros críticos" ["Literary Criticism in Mexico: Our Critics"] 375–381.

35. The original: "Guiados por un principio altamente espiritual y noble, animados de un deseo patriótico, social y literario, puesta la mira en elevados fines, alzamos nuestra humilde y débil voz en defensa de la poesía sentimental, tantas veces hollada, tantas veces combatida, pero triunfante de las desconsoladoras teorías del realismo, y del asqueroso y repugnante positivismo."

36. The original: "¡Y cómo no había de ser así, si nosotros, hijos de la ardiente América, soldados valerosos de la libertad, nacidos en los hermosos valles donde la primavera tiene su dominio eterno, aborrecemos todas las servidumbres, quebrantamos todas las cadenas, y amando con amor infinito todo lo verdadero, lo bueno y lo bello, dejamos volar libremente nuestra imaginación y damos libre curso a todos nuestros nobles sentimientos!"

37. The original: ". . . de manera que hallar imágenes nuevas y hermosas, expresándolas con claridad y concisión, es enriquecer el idioma, renovándolo a la vez. Los encargados de esta obra, tan honorable, por lo menos, como la de refinar los ganados o administrar la renta pública, puesto que se trata de una función social, son los poetas. El idioma es un bien social, y hasta el elemento más sólido de las nacionalidades."

38. The original: "Para decir las nuevas cosas que vemos y sentimos no teníamos vocablos; los hemos buscado en todos los diccionarios, los hemos tomado, cuando los había, y cuando no, los hemos creado."

39. The original: ". . . cuando los *modernistas* usábamos palabras como *aurifabrista,* por orífice; *pucela,* por doncella; *veneficio,* por maleficio, etc., no incurríamos en galicismo alguno, sino que desenterrábamos sencillamente vocablos que habían caído en desuso sin razón, pues, o eran muy bellos, como los dos primeros, o no tenían substitución exacta, como el último."

40. For the background to this section I am indebted to Emir Rodríguez Monegal's illuminating study *El otro Andrés Bello.*

41. The original: "Su tesis, de estirpe romántica, es que un idioma es la expresión de las ideas de un pueblo y un pueblo ha de tomar sus ideas donde ellas estén, independientemente del criterio de pureza idiomática o de perfección académica; que la literatura española ha perdido toda su fuerza y que América ya no está dispuesta a esperar que la mercadería ideológica extranjera pase por cabezas españolas para poder consumirla; que la función real de la Academia Española es recoger, como en un armario, las palabras que usan pueblo y poetas y no *autorizar* el uso de las mismas; que las lenguas vuelven hoy al pueblo (tesis del primer artículo); que el influjo de los gramáticos, el temor de las reglas, el respeto a los admirables modelos, tiene agarrotada la imaginación de los chilenos."

42. The original: "Pero se puede ensanchar el lenguaje, se puede enriquecerlo, se puede acomodarlo a todas las exigencias de la sociedad, y aun de la moda, que ejerce un imperio incontestable sobre la literatura, sin adulterarlo, sin viciar sus construcciones, y sin hacer violencia a su genio" (qtd. in Kristal, n. 19).

43. For further elaboration of this point see Rodríguez Monegal, 317.

44. The original: "La labor del 'pobre narrador americano' acaso resultara 'indisciplinada' o 'informe' (atributos de la barbarie). Pero esa 'espontaneidad', esa cercanía a la vida, ese discurso 'inmediato' era necesario para representar el 'mundo nuevo' que el saber europeo, a pesar de sus propios intereses, desconocía. [. . .] para Sarmiento había que conocer toda esa zona de la vida americana—la barbarie—que resultaba *irrepresentable* para la 'ciencia' y 'los documentos oficiales'. Había que oír al *otro;* oír su voz, ya que el otro carecía de escritura. Eso es lo que el saber disciplinado, y sus importadores, no habían logrado hacer; el *otro* saber—saber del *otro*—resultaría así decisivo en la restauración del orden y del proyecto modernizador."

45. The original: "Sarmiento insistió precisamente en la formación extrauniversitaria de su discurso, espontáneo y hasta indisciplinado, pero por eso más capacitado para entender la 'barbarie' americana." Carlos Fuentes, in his recent novel *La campaña,* represents these struggles for the soul of Spanish America through the protagonist's confrontation with, above all else, the gaucho.

46. For Darío's embrace of Sarmiento's ideas see González-Rodas's "Presencia de Sarmiento en Rubén Darío."

CHAPTER THREE

1. The lesser-studied contemporaries of the four principal early *modernistas* include Manuel González Prada, Justo Sierra, Salvador Díaz Mirón, and Francisco

Gavidia. Though Manuel González Prada (Peru, 1848–1918) is generally more known for his prose than for his poetry, he published nine important volumes that set Spanish American verse on the road toward *modernismo*. Part of his poetry was intellectual and didactic, part was amorous and sentimental, but above all else, it was inventive. After reading Parnassians and symbolists, he wrote imitations, adaptations, and translations; he experimented with their views of language and their formal innovations; he adapted, to Spanish verse, forms from French, English, and Italian; and he incorporated synaesthesia into his work. Justo Sierra (Mexico, 1848–1912) also is known less for his poetry than for his work as a historian, educator, orator, and politician, yet in his verse there are clear pre-*modernista* or early *modernista* elements. His poetry, which is graceful, fresh, elegant, is responsive to the new tendencies that came from abroad, finding inspiration in the Parnassians, Bécquer, D'Annunzio, and Nietzsche, to name just a few.

In contrast to González Prada and Sierra, Salvador Díaz Mirón (Mexico, 1853–1928) stands out for his poetic production. From 1886, when he published his early verse, his presence was felt among *modernista* writers such as Darío, who praised his dynamic, freedom-loving poetry in one of the "Medallones" of the 1890 edition of *Azul . . . .* While Díaz Mirón's tone was revolutionary and his focus was on the human struggle in the urban industrialized centers of Spanish America, he wrote with a grace, learning, and attention to style that appealed even to those concerned primarily with the aesthetics of *modernista* innovations. During his second period, which began with *Lascas* (1901), his revolutionary concerns were directed toward the formal aspects of his poetry. He experimented with musical effects, with accentual and rhythmic changes, and sought, at the expense of the energy that Darío had singled out, a delicate, formal perfection. Like Díaz Mirón, Francisco Gavidia (El Salvador, 1863–1955) stands out for his reform of Spanish American poetic form. He is perhaps best remembered today as one of Darío's teachers and as the poet who introduced into *modernismo* the new rhythms of the French alexandrine and the Greek hexameter.

2. In her analysis of the poem, Eliana S. Rivero explains that Gutiérrez Nájera denied knowing, when he first took the pseudonym, that *Le Duc Job* was a work by the French playwright Léo Laye. He attributed the name, instead, to Manuel Tamayo y Baus and claimed to have been attracted to its "romantic" sound. In the same article, Rivero discusses the multiple references to the foreign features that filled the daily life of high society in Mexico City toward the end of the nineteenth century.

3. Iván A. Schulman and Manuel Pedro González were instrumental in crediting Martí with initiating many of the key characteristics of the movement (Schulman, *Génesis del modernismo;* Schulman and González, *Martí, Darío y el modernismo;* as well as numerous articles). As a result, the beginning of the movement is now often dated by the appearance of *Ismaelillo,* Martí's first published collection, in 1882.

4. See studies by Angel Rama (*Los poetas modernistas en el mercado económico*), Julio Ramos, Iván Schulman (*Relecturas martianas*), Iris Zavala (*Colonialism and Culture*), Blanca Rivera Meléndez, and my contribution to *The Cambridge History of Latin American Literature.*

5. Saúl Yurkievich has written many insightful studies on the relationship between *modernismo* and the avant-garde. Among them, his *Celebración del modernismo,* his "Rubén Darío y la modernidad," and his "Rubén Darío, precursor de la vanguardia" stand out. See also Cintio Vitier, "Vallejo y Martí."

6. The original reads: "Estos son mis versos. Son como son. A nadie los pedí prestados. Mientras no pude encerrar íntegras mis visiones en una forma adecuada a ellas, dejé volar mis visiones: oh, cuánto áureo amigo, que ya nunca ha vuelto! Pero la poesía tiene su honradez, y yo he querido siempre ser honrado. Recortar versos, también sé, pero no quiero. Así como cada hombre trae su fisonomía, cada inspiración trae su lenguaje."

7. The *seguidilla* is a poem of seven lines consisting of a quatrain and a tercet.

8. See Roberto González Echevarría's outstanding article "Martí y su 'Amor de ciudad grande,'" which focuses on this point.

9. The *redondilla* is a four-line stanza whose rhyme scheme is *abba.*

10. My literal prose translation seeks to clarify certain images obscured by the more poetic translation by Elinor Randall published in *José Martí, Major Poems: A Bilingual Edition:*

I know about Persia and Xenophon,
Egypt and the Sudan,
But I prefer to be caressed
By fresh mountain air.
I know the age-old history
Of human grudges,
But I prefer the bees that fly
Among the bellflowers.
I know the songs that breezes sing
In the chattering branches;
Don't tell me that I lie—
I do prefer them.
I know about the frightened buck
Returned to its pen, expiring;
I know that weary hearts die darkly
But free from anger. (65)

11. José Martí, "Congreso internacional de Washington," in *Obras completas,* vol. 6, 46–57. For additional commentary on this article see Roberto Fernández Retamar, *Introducción a José Martí,* pp. 23–26.

12. While most biographers continue to view Silva's death as a suicide, Enrique Santos Molano has suggested the possibility of a conspiracy and murder. For a less radical perspective, see Ricardo Cano Gaviria's *José Asunción Silva: Una vida en clave de sombra.*

13. For the latest views regarding these groupings and the dating of Silva's poetry, see Héctor Orjuela's study and Jesús Munárriz's comments, both in the Hiperión edition of the *Obra poética* (298–299, 303–308).

14. The original: ". . . la novela de Silva es excepcionalmente importante para el estudio de la historia intelectual del modernismo. Viene a ser, a nuestro juicio, uno de los mejores documentos que tenemos para conocer no sólo la crisis personal de un poeta sino también el ambiente intelectualizado, internacional y de salón, en que se movían al menos en sus sueños íntimos los escritores de aquellos años. *De sobremesa* es, pues, clave y testimonio de toda una época."

15. Alfredo Villanueva-Collado summarizes this early critical judgment citing the works of Baldomero Sanín Cano, Juan Loveluck, and Ferdinand Contino ("*De sobremesa,*" 9).

16. Included in these are the studies by Juan Loveluck and Edgar O'Hara.

17. The original: "Recogida por la pantalla de gasa y encajes, la claridad tibia de la lámpara caía en círculo sobre el terciopelo carmesí de la carpeta, y al iluminar de lleno tres tazas de China, doradas en el fondo por un resto de café espeso, y un frasco de cristal tallado, lleno de licor trasparente entre el cual brillaban partículas de oro, dejaba ahogado en una penumbra de sombría púrpura, producida por el tono de la alfombra, los tapices y las colgaduras, el resto de la estancia silenciosa.

En el fondo de ella, atenuada por diminutas pantallas de rojiza gasa, luchaba con la semioscuridad circunvecina la luz de las bujías del piano, en cuyo teclado abierto oponía su blancura brillante el marfil al negro mate del ébano.

Sobre el rojo de la pared, cubierta con opaco tapiz de lana, brillaban las cinceladuras de los puños y el acero terso de las hojas de dos espadas cruzadas en panoplia sobre una rodela, y destacándose del fondo oscuro del lienzo, limitado por el oro de un marco florentino, sonreía con expresión bonachona la cabeza de un burgomaestre flamenco, copiada de Rembrandt."

18. Several critical studies have focused on the relationship between *De sobremesa* and decadentism—especially the impact of *A Rebours* by Huysmans. One of the most extensive works is by Klaus Meyer-Minnemann.

19. Juan De Garganta examines Silva's politics relying on lengthy sections of *De sobremesa.* I can agree with little in his study, for he finds in *De sobremesa* only "documentary value."

20. Robert Jay Glickman's article "José Asunción Silva ante los avances tecnológicos de su época" provides interesting insights into this situation.

21. The original: "Como se ve, la problemática fundamental del decadentismo literario era la cuestión acerca de los límites, los linderos, de la literatura: los numerosos «ismos» en las artes plásticas y en las letras del fin de siglo y principios del siglo XX . . . son sintomáticos del ansia que entonces sentían los artistas por definir y delimitar la naturaleza de su quehacer."

22. The original: "Esa meditación acerca de intelectual, en *De sobremesa,* se centra particularmente . . . en torno a la necesidad de definir las fronteras entre el mundo de las ideas—de los textos, de las ficciones—que maneja el intelectual, y el mundo de las acciones concretas."

23. The original: "un marco transcendente, un límite absoluto que le sirva de punto de referencia y que justifique los demás marcos."

24. The original: "imagen del escritor finisecular como un ser que opta

deliberadamente por dedicarse a la producción de ficciones y a la contemplación idolátrica de éstas."

25. The original: "una novela hermética, que a través de símbolos alquímicos y rosacruces describe una vía de purificación espiritual a la cual el protagonista, José Fernández, se somete bajo la influencia de su amor por Helena D'Scilly quien, junto a su abuela, representa el principio femenino regenerador en la novela."

26. It should not be necessary to clarify that the Nietzschean features that appear in the novel correspond to a popular conception of the German's philosophy. Aníbal González has correctly pointed out that: ". . . un aspecto interesante de la novela de Silva es que testimonia el temprano (aunque superficial) conocimiento que tenía Silva de la obra de Nietzsche en 1896, cuando aún no se había traducido ningún libro del filósofo alemán al castellano, y las noticias sobre su obra, tanto en el mundo hispánico como en Francia e Inglaterra, eran escasas" (*La novela modernista* 105) [". . . an interesting aspect of Silva's novel is that it bears witness to the early (though superficial) knowledge that Silva had of Nietzsche's work in 1896, when no book of the German philosopher had yet been translated, and news about his work, in the Hispanic world as much as in France and England, was scarce"]. What is important, therefore, is Silva's grasp and articulation of the Nietzschean concepts with which he became acquainted. Because of their general nature, the extent to which Silva's rendition approximates Nietzsche's overall premises is remarkable.

27. The original: "Ayer saltó otro edificio destrozado por una bomba explosiva, y la concurrencia mundana aplaudió en un teatro del *boulevard,* hasta lastimarse las manos, *La casa de muñecas* de Ibsen, una comedia al modo nuevo en que la heroína, Nora, una mujercilla común y corriente, con una [*sic*] alma de eso que se usa, abandona marido, hijos y relaciones para ir a cumplir los deberes que tiene consigo misma, con un yo que no conoce y que se siente nacer en una noche como hongo que brota y crece en breve espacio de tiempo. Así, a estallidos de melinita en las bases de los palacios y a golpes de zapa en lo más profundo de sus cimientos morales, que eran las antiguas creencias, marcha la humanidad hacia el reino ideal de la justicia que creyó Renán entrever en el fin de los tiempos."

28. The original: "Ibsen y Ravachol le ayudan, cada cual a su modo; cae el primer magistrado de Francia herido por el puñal de Cesáreo Santo, y escribe Suderman *La dama vestida de gris,* donde la abnegación y el amor a la familia toman tintes de sentimientos grotescos, sin que el final de cuento de hadas, agregado por el novelista a su obra como un farmaceuta hábil echaría jarabe para dulcificar una pócima que contuviera estricnina, alcance a disimular el acre sabor de la letánica droga."

29. The original: "Oye, obrero, que pasas tu vida doblado en dos, cuyos músculos se empobrecen con el rudo trabajo y la alimentación deficiente, pero cuyas encallecidas manos hacen todavía la señal de la cruz, obrero que doblas la rodilla para pedirle al cielo por los dueños de la fábrica donde te envenenas con los vapores de las mezclas explosivas, oye, obrero, ¿nada evocan en tu rudimentario cerebro las rudas sílabas de ese nombre germano, Nietzsche, cuando vibran en tus oídos? . . ."

30. The original: "Es que la humanidad había estado recibiendo como ver-

daderas nociones falsas sobre su origen y su destino, y el profundo filósofo encontró una piedra de toque en qué ensayar las ideas como se ensayan las monedas para saber el oro que contienen. Eso es lo que se llama reavaluar todos los valores."

31. The original: "Si la conciencia son las garras con que te lastimas y con que puedes destrozar lo que se te presente y coger tu parte de botín en la victoria, no te las hundas en la carne, vuélvelas hacia afuera; sé el sobrehombre; el *Übermensch* libre de todo prejuicio, y con las encallecidas manos con que haces todavía, estúpido, la señal de la cruz, recoge un poco de las mezclas explosivas que te envenenan al respirar sus vapores y haz que salte en pedazos, al estallido del fulminante picrato, la fastuosa vivienda del rico que te explota. Muertos los amos, serán los esclavos los dueños y profesarán la moral verdadera en que son virtudes la lujuria, el asesinato y la violencia. ¿Entiendes, obrero? . . ."

32. The original: ". . . Mira: del oscuro fondo del Oriente, patria de los dioses, vuelven el budismo y la magia a reconquistar el mundo occidental. París, la metrópoli, les abre sus puertas como las abrió Roma a los cultos de Mitra y de Isis; hay cincuenta centros teosóficos, centenares de sociedades que investigan los misteriosos fenómenos psíquicos; abandona Tolstoi el arte para hacer propaganda práctica de caridad y de altruísmo, ¡la humanidad está salvada, la nueva fe enciende sus antorchas para alumbrarle el camino tenebroso!"

33. The original: "¡Helena! ¡Helena! ¡Me corre fuego por las venas y mi alma se olvida de la tierra cuando pienso en esas horas que llegarán si logro encontrarte y unir tu vida con la mía! . . ."

34. The original: "Una oleada poderosa de sensualismo me corre por todo el cuerpo, enciende mi sangre, entona mis músculos . . ."

CHAPTER FOUR

1. The original: "El Modernismo que ahora me interesa—no niego que haya otros—es el que nace y muere con Rubén Darío."

2. As already noted in chapter 2, the deliberate transgressing of literary, social, and sexual limitations is also essential to Silva's novel *De sobremesa,* which was rewritten, after having been lost in a shipwreck, in 1896.

3. The original: "Sin pinceles, sin paleta, sin papel, sin lápiz, Ricardo, poeta lírico incorregible, huyendo de las agitaciones y turbulencias, de las máquinas y de los fardos, del ruido monótono de los tranvías y el chocar de los caballos con su repiqueteo de caracoles sobre las piedras; del tropel de los comerciantes; del grito de los vendedores de diarios; del incesante bullicio e inacabable hervor de este puerto; en busca de impresiones y de cuadros, subió al Cerro Alegre, que, gallardo como una gran roca florecida, luce sus flancos verdes, sus montículos coronados de casas risueñas escalonadas en la altura, rodeadas de jardines, con ondeantes cortinas de enredaderas, jaulas de pájaros, jarras de flores, rejas vistosas y niños rubios de caras angélicas."

4. Darío himself comments: "Mas el azul era para mí el color del ensueño,

el color del arte, un color helénico y homérico, color oceánico y firmamental, el «coeruleum», que en Plinio es el color simple que semeja al de los cielos y al zafiro" ["But blue was for me the color of enchantment, the color of art, an Hellenic and Homeric color, oceanic and heavenly color, the 'coeruleum' that in Pliny is the simple color that approaches that of the sky and sapphires"] (*Obras completas* 1: 197). Valera assumes that the title alludes to Hugo's *L'art, c'est l'azur,* but Raymond Skyrme, in an informative article on the subject, shows that the Huysmans's *A Rebours* is more likely to be the immediate source of the title ("Darío's *Azul* . . .").

5. Rubén Benítez, while focusing on the poem's artistic elements, recognized its political implications. He suggests that "Estival" proposes a positive reevaluation of those aspects of the world and of man that bourgeois civilization rejects as barbaric.

6. Works that have examined Darío's "erotic mysticism" are Paz's "El caracol y la sirena," Guillermo Díaz-Plaja's *Modernismo frente a noventa y ocho,* Graciela Palau de Nemes's "Tres momentos del neomisticismo poético del 'siglo modernista': Darío, Jiménez y Paz," Pedro Salinas's *La poesía de Rubén Darío,* and Skyrme's *Rubén Darío and the Pythagorean Tradition.* The critic who stands out for positing a political dimension to this search for transcendence is, as noted in chapter 1, Octavio Paz (*Los hijos* 54–60).

7. The esoteric religions in vogue at the time provided a legitimizing framework for this orientation.

8. For a more extensive analysis of this poem see Keith Ellis, *Critical Approaches to Rubén Darío,* 90–95.

9. For further details about Sawa, see Allen Phillips's revealing study. Iris Zavala's introduction to Sawa's *Iluminaciones en la sombra* sheds light on the literary, social, and political trends in Spain at the turn of the century.

10. I call "Palabras liminares" a nonmanifesto manifesto because Darío insists: ". . . solicitaron lo que, en conciencia, no he creído fructuoso ni oportuno: un manifiesto" [". . . they requested what, in good conscience, I did not believe to be fruitful or opportune: a manifesto"] (*Poesía* 179). For more about Darío's "manifestos," see Saúl Yurkievich's "El efecto manifestario, una clave de modernidad."

11. The original: "Yo he dicho, en la misa rosa de mi juventud, mis antífonas, mis secuencias, mis profanas prosas . . . Tocad, campanas de oro, campanas de plata; tocad todos los días, llamándome a la fiesta en que brillan los ojos de fuego, y las rosas de las bocas sangran delicias únicas."

12. The original: "Como cada palabra tiene un alma, hay en cada verso, además de la harmonía verbal, una melodía ideal. La música es sólo de la idea, muchas veces."

13. The original: "Y la primera ley, creador: crear. Bufe el eunuco. Cuando una musa te dé un hijo, queden las otras ocho encinta."

14. The reference is, of course, to Verlaine's 1869 collection of verse, *Fêtes Galantes.*

15. Ricardo J. Kaliman also recognizes the importance of "Era un aire suave . . ." as a *modernista* manifesto, emphasizing the parallel between Eulalia and

poetic language. His focus, however, is on the tension between the Parnassian and symbolist strains in *modernista* verse. He believes that, in the same way that Eulalia rebels against the aristocratic and refined flirtations of her suitors, preferring the authenticity of the flesh (the page), the *modernista* poet subverts the Parnassian model through the restoration of sexual desire as the factor that defines the image (31).

16. Confusion regarding the nature of "Divagación" has stemmed from the stubborn insistence on the part of critics to read it as a love poem. Alberto J. Carlos begins his erudite article with Salinas's statement on the preeminence of "lo amoroso" in *Azul . . .* and *Prosas profanas.* He continues his detailed study along these lines, concluding with regard to the initial "invitada" ["invited one"] that "No hay duda de que se trata de Venus" ["There is no doubt that it is about Venus"] (293, 296). The internal inconsistencies that arise out of this assumption are confronted only in passing. Later Alonso Zamora Vicente, starting with the same premise as Carlos and Salinas, states the problem directly: "Se trata de una poesía hecha de componentes de cultura, históricos, artísticos, etc. Pero, ¿dónde está el sentimiento amoroso? No aparece por ninguna parte" ["It is a question of a poetry made of cultural, historical, artistic components. But where is the amorous sentiment? It does not appear anywhere"] (173).

17. For detailed analyses of the references that appear throughout the poem, see the articles by Carlos and Zamora Vicente.

18. For the relationship of the museum to philology and the role of both in *modernista* literature, see Aníbal González's *La crónica modernista,* 22–23, 29.

19. Darío's concern regarding his ability to step out of his cultural context and to assume the role he has set for himself is alluded to in the unexpected and, perhaps, ironic dating of the poem. Following the last line of verse, he includes the information "Tigre Hotel, diciembre 1894," as if to direct attention to the difficulty of speaking with a universal voice while locked within a specific and restrictive milieu. The juxtaposition of the mystical conclusion of the poem with a concrete time and place—one associated with the comfortable bourgeoisie—points to the issues raised by the poem itself.

20. See John Dyson and Alberto Forcadas for additional background to "Sonatina."

21. In "Mientras tenéis, oh negros corazones . . ." (*Poesía* 258), the concluding apocalyptic vision consoles the suffering poet, who is converted into a Christlike figure. He leads his readers into an ideal future on the back of Pegasus.

22. Much of the critical attention directed toward "Sonatina" as well as "Era un aire suave . . ." has focused on the musicality referred to in their titles. Critics have examined the presence of musical forms, instruments, and imagery in "Era un aire suave . . . ," the musical structure of "Sonatina," and the inner musicality of both. Nevertheless, the role of music as a central poetic ideal linked to a transcendent, pure, untarnished vision of the universe has tended to be overlooked. See Enrique Anderson Imbert, 78–81; Helmut Hatzfeld; and Tomás Navarro

Tomás. Pat O'Brien tends to trivialize the entire endeavor by continuing to read "Sonatina" as a manifesto of escapist poetry.

23. As part of his phonological analysis, Navarro Tomás points out the "prestigio señorial" ["lordly prestige"] of the nouns (198).

24. Though I have studied this poem extensively in terms of its recourse to esoteric tradition, here I seek to examine it with regard to *modernista* aspirations (*Rubén Darío*, 27–45).

25. The best article on this poem and on the importance of the sea within Darío's poetic universe remains the one written by my erudite friend and teacher Alan S. Trueblood. See also Francisco Sánchez-Casteñer's *Rubén Darío y el mar.*

26. The original: "Si en estos cantos hay política, es porque aparece universal. Y si encontráis versos a un presidente, es porque son un clamor continental. Mañana podremos ser yanquis (y es lo más probable); de todas maneras, mi protesta queda escrita sobre las alas de los inmaculados cisnes, tan ilustres como Júpiter."

27. Michael P. Predmore presents a beautifully executed stylistic analysis of "Lo fatal," which complements Anderson Imbert's brilliant examination in his *La originalidad de Rubén Darío* (141–147).

28. The original: "El clisé verbal es dañoso porque encierra en sí el clisé mental, y, juntos, perpetúan la anquilosis, la inmovilidad."

29. The original: "El verdadero artista comprende todas las maneras y halla la belleza bajo todas las formas. Toda la gloria y toda la eternidad están en nuestra conciencia."

30. Darío experiments with form throughout *El canto errante.* There is a brilliantly executed *eco* ("Eco y yo" ["Echo and I"]), an extensive and revealing *epístola* to Mrs. Leopoldo Lugones ("Epístola" ["Letter"]), and a ten-part ode written in memory of Bartolomé Mitre ("Oda" ["Ode"]). There are poems about America ("A Colón" ["To Columbus"], "Momotombo" ["Momotombo"], "Desde la Pampa" ["From the Pampas"], "Tutecotzimí" ["Tutecotzimí"]), including one in praise of the United States ("Salutación al águila" ["Greetings to the Eagle"]), poems that evoke the ancient Mediterranean ("Revelación" ["Revelation"], "Hondas" ["Slingshots"], "Eheu!" ["Alas!"], "La canción de los pinos" ["The Song of the Pines"]), and others that portray the magical worlds of art, fantasy, and self-indulgence ("A Francia" ["To France"], "Visión" ["Vision"], "La hembra del pavo real" ["The Female of the Peacock"], "Danza elefantina" ["Elephantine Dance"], "La bailarina de los pies desnudos" ["The Barefoot Ballerina"], "Dream," "Balada en honor de las musas de carne y hueso" ["Ballad in Honor of the Muses of Flesh and Blood"], "Flirt"). There are also poems about poets and poetry ("Antonio Machado," "Preludio" ["Prelude"], "Campoamor," "Soneto" ["Sonnet"]) and poems that offer reincarnation as an alternative to the orthodox view of human destiny ("Eheu!," "Hondas," "Metempsícosis" ["Metempsychosis"]). As in the first and title poem of the collection, the images come from all corners of the globe, Greek and Roman mythology, the Bible, world literature, and modern life. Despite this syncretic vision and the many bases for consolation and optimism that it offers, Darío is unable to conquer completely the profound anguish and despair that

became particularly acute with advancing age and the approach of death ("Sum," "Eheu!," "Nocturno," "Epístola").

CHAPTER FIVE

1. The original: "... todos los poetas de este momento intermediario aceptaron el modernismo como punto de partida, para después, siguiendo cada uno de ellos su propio criterio, modificarlo o rechazarlo. Algunos nunca rompen del todo con la temática y las formas anteriores; otros van rápidamente hacia los experimentos del vanguardismo o hacia las realidades americanas del criollismo."

2. The original: "En realidad el poema no era ... sino la expresión reactiva contra ciertos tópicos modernistas arrancados al opulento bagaje lírico de Rubén Darío, el Darío de *Prosas profanas* y no el de *Cantos de vida y esperanza.* Dejando a un lado lo esencial en la poesía del gran nicaragüense, se prolongaba en sus imitadores lo que podríamos llamar exterioridad y procedimiento. Claro está que en los imitadores faltaban la gracia, el virtuosismo excepcional y la encantadora personalidad del modelo. No alcanzaban tampoco los secuaces de Darío su emoción lírica, perceptible en él desde *Prosas profanas,* aun en poemas donde la agilidad técnica y el dominio de la forma parecían la única intención creadora; mucho menos la que, en *Cantos de vida y esperanza,* lograra, ya íntegra, madura y sabia, la poesía de Rubén. Lo único que estaba a la mano de los imitadores era lo temático—cisnes, pajes, princesas—; la métrica—ya tomada de Francia o de la vieja poesía española—; la adjetivación, que a fuerza de repetida por ellos perdía eficacia y novedad; en general, la palabra, estéril para quien la hurta, y no el espíritu, fecundo y renovador."

3. Included among these poets were Baudelaire, Heredia, Verlaine, Maeterlinck, Jammes, and Rodenbach.

4. For a more complete analysis of the way these concepts are developed in Nervo's poetry, see my "El Modernismo y la Generación del '98."

5. Joseph F. Vélez's analysis of this poem coincides to a large extent with mine, although it overlooks the implied critique of social values derived from pragmatism and materialism.

6. See, for example, "Nueva escuela literaria," *Obras completas,* 2: 178–182.

7. See José Emilio Pacheco's footnote to the poem, *Antología* 2: 27–28.

8. For further background information see Manuel Durán, *Genio y figura de Amado Nervo,* and Luis Leal, "Situación de Amado Nervo."

9. Jaimes Freyre's practice of free verse began within the literary circle headed by Darío and Lugones between 1893 and 1898, working, as he stated, a little by intuition and a little under the influence of French, Italian, and Portuguese writers. His first collection of poems, *Castalia bárbara* [*Barbaric Castalia*], was published in Buenos Aires in 1899 and contains six of the earliest examples of free verse in Spanish poetry. They divide into two types, one based on the *silva* (a poem that consists of lines of seven and eleven syllables with no fixed pattern or stanza length) and the other on prosodic groupings. His second and only other collection of poetry, *Los*

sueños son vida [*Dreams Are Life*] (1917), also contains three poems of these two types along with "Alma helénica" ["Hellenic Soul"], a complex, polymetric poem.

Between the publication of these two collections of verse, he published *Leyes de versificación castellana* [*The Rules of Castilian Versification*] in 1912, in which he spelled out his basic assumptions and beliefs about Spanish poetics. In this work, he held that only accent has the ability to generate rhythm and that Spanish verse forms are created by combining prosodic groupings (*períodos prosódicos*). This idea of verse as prosodic groupings allowed him to distinguish between prose and poetry: unlike the former, the latter is characterized by a combination of equal or analogous groupings. The emphasis upon accentual rhythm, instead of upon the mechanical counting of syllables, provided theoretical support for the type of free verse that he had already begun to write and resuscitates, as noted by Henríquez Ureña, the irregular versification of old Spanish verse (179–183). Despite precedents in the distant past, Jaimes Freyre claims for himself the honor of having been the first to introduce free verse into Spanish. He gives 1894 as the date of his breakthrough, thereby making it contemporaneous with Silva's "Nocturno," which, written in 1892 and published in 1894, expanded Spanish metrics through the free accumulation of clauses. For additional information, see Isabel Paraíso de Leal, "Teoría y práctica del verso libre en Ricardo Jaimes Freyre."

10. The original: "rebeldía originado al contacto con la realidad mezquina . . . rebeldía contra el destino del hombre, no solamente condenado a morir, sino a vivir en sociedades regidas por el materialismo más crudo."

11. The original: "da lo mismo refugiarse en las brumas del Norte—como Ricardo Jaimes Freyre—, en el Versalles rubeniano o en las chinoserías de Julián del Casal."

12. Enrique Anderson Imbert suggests that Wagner's lyrical dramas were the source of Jaimes Freyre's references (1: 417). But acquaintance with Wagner's ideas and ideals may have also been indirect, through the works of others. As Raymond Furness has shown, Wagner was one of the most influential artists of his time. He had a profound impact on the most outstanding figures of the day, such as Baudelaire, Mallarmé, Huysmans, Verlaine, and Rimbaud. Giovanni Allegra also traces the influence of Wagner in the shaping of what Juan Ramón Jiménez calls the "*modernista* mentality." He focuses on the works of López Chavarri and Bonilla y San Martín.

13. Written in 1893, one alludes to *Lohengrin* and the other to *Parsifal.* These sonnets do not appear in the Ayacucho edition. They can be found in the edition of Darío's *Poesías completas* compiled by Alonso Méndez Plancarte and published by Aguilar in Madrid in 1967 (963–964).

14. Mireya Jaimes Freyre studies this phenomenon but ignores completely its philosophic dimension. She focuses, instead, on "la cualidad pictórica" ["the pictorial characteristic"] of the verse. Since she follows the early—and largely erroneous—critical dichotomy between *modernismo* and the Generation of '98, she finds in "El hospitalario" ["The Hospitaler"] "una actitud digna de un personaje de una obra de los del 98" ["an attitude worthy of a character in a Generation of

'98 work"]. On the contrary, the lone fighter who struggles against the horrors of anguish, death, and selfishness is just as likely to be the poet-visionary-savior already observed in so many *modernista* works.

15. For a detailed structural analysis of this poem see David William Foster's article.

16. For an analysis of Wagner's *The Ring of the Nibelung*, see Alan David Aberbach's *The Ideas of Richard Wagner* (293–431).

17. On Valencia as translator, see Alan Trueblood, "Wilde y Valencia: *La balada de la Cárcel de Reading*."

18. Although Mariátegui makes some very astute observations about Eguren's poetry, he insists on what he perceives to be the artist's inability to accept his own reality.

19. Rodríguez-Peralta acknowledges that many critics have chosen to focus on those aspects of Eguren's work that depart from *modernista* tendencies. She refers to Monguió's *La poesía postmodernista peruana*, Anderson Imbert's *Historia de la literatura hispanoamericana*, and Estuardo Núñez's *José María Eguren: Vida y obra*. However, her article seeks to show the *modernista* side of his endeavors. She observes that his poetry is filled with exoticism, oriental fantasies, and references to both classical and Nordic mythologies; his vocabulary includes neologisms, French, and Italianate forms. There are also examples of the frivolous, elegant, and aristocratic details that for many defined much of early *modernismo*.

20. The original: "Sus amigos y sus críticos vieron en las escenas infantiles de esta poesía el cándido poeta que parecía jugar. No vieron otra cosa: el horror de ese juego, la mirada del poeta asaltada por el pavor, la vida y la muerte que Eguren quiso mostrar en toda su fragilidad en una reiterada escena infantil."

21. Here I am indebted to Roberto González Echevarría's introduction to his *Voice of the Masters* for his lucid formulation of the role of literature in Spanish America.

CHAPTER SIX

1. The original: "un hombre que, sin saberlo, se negó a la pasión y laboriosamente erigió altos e ilustres edificios verbales hasta que el frío y la soledad lo alcanzaron. Entonces, aquel hombre, señor de todas las palabras y de todas las pompas de la palabra, sintió en la entraña que la realidad no es verbal y puede ser incomunicable y atroz, fue callado y solo a buscar, en el crepúsculo de una isla, la muerte."

2. Gwen Kirkpatrick, I believe, overstates the case. Her exaggeration sets up a special place for Lugones and Herrera y Reissig. She has them breaking out of conditions that, in reality, were already in flux. She writes: "Much of what seems tedious in *modernista* poetry for the modern reader is its overloading of rarefied objects, its jewel-studded interior spaces, the amethyst shafts of light that make vision difficult. We find it hard to move around these ornately furnished rooms

and especially amidst the heavy-lidded goddesses who inhabit them. While modern taste prefers clean, spare lines, white walls, and open spaces, the *modernistas* work from a different set of culturally determined preferences. Just as they held a penchant for ornately decorated physical spaces, language itself had to be filled, decorated, and overburdened until it groaned under the excess of sensory paraphernalia. With rhyme, rhythm, and extended imagistic development, every inch of space was filled, inviting crowding, violence, and ultimately, parody. And this is precisely the process we see in several late *modernista* poets. Growing agitation, slicing through not only the images but the very contours of the poems themselves, carried *modernista* innovation to frenzies of linguistic activity" (*Dissonant Legacy* 6–7).

3. For valuable background to Lugones's early literary and political ideas, see Kirkpatrick's study of Lugones's journalistic writings published between 1893 and 1898, "Art and Politics in Lugones's Early Journalism."

4. Lugones's faith in the harmony of the universe was reinforced by his interest in the occult sciences, in general, and theosophy, in particular. Studies dealing with this aspect of Lugones's artistic creation include Hewitt and Hall, Marini Palmieri, Dmitrowicz, and Fraser.

5. This reference is found in Darío's article "Leopoldo Lugones," in *Semblanzas*, in volume 2 of his *Obras completas*, 990–993.

6. This tendency reappears in the pastoral works of Herrera y Reissig and Ramón López Velarde. Allen Phillips also sees this connection with Herrera. In addition, he finds in these poems a "realist" tendency ("Cuatro poetas" 434). Saúl Yurkievich examines this "realist" tendency more fully, linking these poems to Lugones's later works (*Celebración* 62–63).

7. For a very different analysis of this poem see Víctor J. Rojas.

8. In his analysis, Orlando Rodríguez-Sardiñas focuses on the formal aspects of the poem.

9. The original: "la utilidad del verso en el cultivo de los idiomas . . ."

10. The original: "Siendo conciso y claro, tiende a ser definitivo, agregando a la lengua una nueva expresión proverbial o frase hecha que ahorra tiempo y esfuerzo . . ."

11. The original: "El idioma es un bien social, y hasta el elemento más sólido de las nacionalidades."

12. Even more strongly ironic is Lugones's "Himno a la luna" ["Hymn to the Moon"], also from *Lunario sentimental.* In a brief article on this poem, E. Caracciolo Trejo makes two important observations that reinforce the conclusions that I have drawn. First, "Himno a la luna" picks up a current that exists throughout modernism and that can be observed as early as in Silva's *Gotas amargas.* Second, the poem reflects a harsh critical perspective that had not been seen in Spanish poetry since the seventeenth century.

13. The original: "La extremada sublimación y estilización de Herrera y Reissig responde a un rechazo radical de todo utilitarismo, a una empedernida desafección del orden fundado en el provecho, a una oposición a las formas de vida establecidas. Su poesía actúa como mediación distanciadora de la existencia alie-

nada, quiere recuperar por el extrañamiento la trascendencia inalcanzable en la práctica social. Es la denuncia de una ausencia, de una mutilación, de una dimensión carente. Es arte subversivo: propone una recreación imaginaria de la experiencia fáctica. Negación del orden imperante, negación de todo orden represivo, la poesía de Herrera y Reissig se desconecta por completo de lo circunstancial y circundante para preservar una libertad que sólo puede darse en la dimensión estética. Los valores estéticos, aunque irrealizables, implican la repulsa de los valores dominantes . . . La poesía . . . [s]e refugia en una integridad ilusionista, en un universo de exaltada ficción para oponerlo a la violencia reductora del mundo factible."

14. See, for example, the statement made by Idea Vilariño and Alicia Migdal, "Criterio de esta edición," in the Ayacucho edition of Herrera y Reissig's *Poesía completa y prosa selecta.*

15. See my analysis of the poem in *Rubén Darío and the Romantic Search for Unity,* 45–46, 114–117.

16. For Herrera y Reissig's interest in the Baroque in general and Góngora in particular see Kirkpatrick, 176–177. She correctly indicates that this sense of artistic accord with the earlier movement anticipates trends that show up more emphatically in the Generation of 1927—and, it should also be noted, in the works of the Spanish American Neo-Baroque.

17. Magdalena García Pinto's introduction to Agustini's *Poesías completas* includes an interesting and informative review of the way Agustini's work has been regarded in various literary histories.

CHAPTER SEVEN

1. The Uruguayan *modernista* José Enrique Rodó elaborates upon this linkage between pristine beauty and moral integrity in his classic essay *Ariel.* "Even if love and appreciation of beauty did not respond to some essential need in rational man, or if they were not in themselves deserving of cultivation, a higher morality would dictate a culture of aesthetics simply in the best interests of society. No one is averse to educating the moral sensibility, and preparing the individual to appreciate beauty should be implicit in that education. Never forget that an educated sense of the beautiful is the most effective collaborator in the formation of a delicate sensitivity for justice" (49–50). Gene H. Bell-Villada discusses Rodó and other *modernistas* in his wide-ranging study of the ongoing relationship between aesthetics and social concerns in literature entitled *Art for Art's Sake and Literary Life.*

2. The original: "El arte moderno no sólo es el hijo de la edad crítica sino que también es el crítico de sí mismo" (*Los hijos* 18).

3. Though she tends to view *modernismo* as a movement aligned with and supported by the *status quo,* Jean Franco acknowledges Vallejo's debt to the same romantic defiance that first appeared in Spanish American literature among the *modernistas.* "To shake his fist at the burghers of Trujillo, [Vallejo] had to abandon his family gods and with them the Christian Logos. His experience in the mines

and on the sugar estates showed him that the moral teachings of Christianity were a dead letter; the one counter-ideology available to him was that of the poet-demon, the literary rebel who takes on the whole of material creation. *Los heraldos negros* is imbued with the Romantic myth of man's fall into division and separation, his yearning for wholeness and the circular journey back to the peace of death though it is a myth that the poet accepts only to destroy it" (27).

4. Ricardo Gutiérrez Mouat's study "La presencia de ciertos textos de Darío en *Residencia en la tierra*" provides a revealing example of the strong and immediate impact of *modernista* texts on the poetic goals of later poets.

5. Cortázar's interest in romantic and postromantic literature has been studied by Ana Hernández del Castillo in *Keats, Poe, and the Shaping of Cortázar's Mythopoesis.* She is particularly interested in Cortázar's recourse to archetypes, rites, and mysteries to maintain a mystical dimension to his writing. See also "*Los reyes:* Cortázar's Mythology of Writing" in *The Voice of the Masters* by Roberto González Echevarría.

6. For a more thorough examination of the significance of this quotation, see my article "El significado de un vínculo textual inesperado: *Rayuela* y 'Tuércele el cuello al cisne.'"

7. The original: "a lo mejor en eso precisamente estaba la victoria" (*Rayuela* 240).

8. The original: "con armas fabulosas, no con el alma del Occidente, con el espíritu, esas potencias gastadas por su propia mentira . . ." (*Rayuela* 240).

9. In his defense of his writing against the attack made by Oscar Collazos, Cortázar acknowledges what the *modernistas* understood all too well, that is, that one must comprehend the true nature of existence in all its dimensions in order to be able to make perceptive, just, and worthy political decisions. He states: "Authentic reality is much more than the 'sociohistoric and political context' . . . and therefore a literature that deserves that name is that which strikes man from all angles (and not, because he belongs to the third world, only or primarily from the sociopolitical angle); it is that which elevates him, stimulates him, changes him, justifies him, drives him crazy, makes him more real, more human, like Homer made the Greeks more real, that is, more human, like Martí and Vallejo and Borges made Latin Americans more real, that is to say, more human." [The original: "La auténtica realidad es mucho más que el 'contexto sociohistórico y político' . . . y por eso una literatura que merezca su nombre es aquella que incide en el hombre desde todos los ángulos (y no, por pertenecer al tercer mundo, solamente o principalmente en el ángulo sociopolítico), que lo exalta, lo incita, lo cambia, lo justifica, lo saca de sus casillas, lo hace más realidad, más hombre, como Homero hizo más reales, es decir más hombres, a los griegos, y como Martí y Vallejo y Borges hicieron más reales, es decir más hombres, a los latinoamericanos"] ("Literatura" 65). Here Cortázar places himself in the same trajectory that I have drawn: from Martí to Vallejo to Borges to himself.

10. "«Las cosas tienen vida propia—pregonaba el gitano con áspero acento—todo es cuestión de despertarles el ánima»" (*Cien años* 80). Jacques Joset

in a footnote to this phrase acknowledges that with it García Márquez establishes one of the keys to the novel, its "magical" dimension.

11. Darío's name appears on pages 8 and 267 in the Spanish and on pages 10 and 247 in the English.

12. The original: ". . . durante las dos horas del recital soportamos la certidumbre de que él estaba ahí, sentíamos la presencia invisible que vigilaba nuestro destino para que no fuera alterado por el desorden de la poesía, él regulaba el amor, decidía la intensidad y el término de la muerte en un rincón del palco en penumbra desde donde vio sin ser visto al minotauro espeso cuya voz de centella marina lo sacó en vilo de su sitio y de su instante y lo dejó flotando sin su permiso en el trueno de oro de los claros clarines de los arcos triunfales de Martes y Minervas de una gloria que no era la suya mi general . . ." (*El otoño* 194).

13. The original: ". . . se sintió pobre y minúsculo en el estruendo sísmico de los aplausos que él aprobaba en la sombra pensando madre mía Bendición Alvarado eso sí es un desfile, no las mierdas que me organiza esta gente, sintiéndose disminuido y solo, oprimido por el sopor y los zancudos y las columnas de sapolín de oro y el terciopelo marchito del palco de honor, carajo, cómo es posible que este indio pueda escribir una cosa tan bella con la misma mano con que se limpia el culo, se decía, tan exaltado por la revelación de la belleza escrita . . ." (*El otoño* 194–195).

14. While I have aimed to demonstrate that *modernismo* plays a crucial role in García Márquez's literary imagination and aspirations, Michèle Sarrailh's insistence that Darío and his poetry are a constant and palpable presence in the works of the great Colombian writer appears overstated. She sees the seven stories of *La increíble y triste historia de la cándida Eréndira y de su abuela desalmada* as "an allegorical religious play about the Darío adventure and the *modernista* happening, at the same time as a literary pastiche of Darío's poetic production" ["un auto alegórico de la aventura rubeniana y del acontecimiento modernista, al mismo tiempo que como «pastiche» literario de la creación de Rubén Darío"] ("Rubén Darío" 707). Also, because of Darío's supposed authority over his contemporaries and successors, she identifies Darío with the dictatorial protagonist of *El otoño del patriarca.* She holds that "[o]nly in the very last pages of the book, will the yoke of the father of the American homeland be thrown off, at the same time García Márquez's generation will be able—only as far as one can imagine—to distance themselves to a healthy degree from the Darío phenomenon" ["[s]ólo en las ultimísimas páginas del libro, sacudiráse el yugo del padre de la patria americana, al tiempo que la generación de García Márquez conseguirá—hasta donde uno lo puede creer—distanciarse saludablemente del fenómeno dariano"] ("Apuntes" 435). Despite the established metaphoric link of author with dictator, Sarrailh's position appears extreme.

15. González Echevarría notes: "The question of the artist in contemporary society is, as in *The Lost Steps,* the manifest theme of 'Manhunt,' and this Romantic topic of the artist's alienation is couched in terms of a longing for the absolute, for oneness and restoration. This desire involves once more returning to a lost paradise,

the world of nature and the mother, and a search for God, the maker of order and giver of meaning in the universe. The search takes various general forms: love, religion, art, and political action" (*Alejo Carpentier* 191).

16. The original: ". . . la literatura no sólo mantiene una experiencia histórica dada, no sólo continúa una tradición, sino que, mediante el riesgo moral y la experimentación formal y el humor verbal, rompe el horizonte conservador de los lectores y contribuye a liberarnos a todos de las cadenas de una percepción antigua, de una matriz estéril, de un prejuicio añejo y doctrinado." I am indebted to Anne Petit for bringing this article to my attention. Her insightful dissertation sheds light on the way Carlos Fuentes develops many of these issues.

17. The original: "En la percepción que Darío tiene del mundo como un código significante hay un intermediario. Ese intermediario es *siempre* de orden cultural, es decir, que Darío introduce en la literatura esta dimensión fundamental . . . : el poema se sitúa en una esfera absolutamente cultural, en lo que los estructuralistas llaman 'el código de papel.' Este intermediario, siempre plástico en él, es con frecuencia también de orden literario, y cuando digo literario, digo Verlaine."

18. My thinking here has been influenced by González Echevarría's incomparable studies on Sarduy and by the brilliant dissertation and writings by Ana Eire.

Works Cited

Aberbach, Alan David. *The Ideas of Richard Wagner: An Examination and Analysis of His Major Aesthetic, Political, Economic, Social, and Religious Thoughts.* Rev. ed. Boston: University Press of America, 1988.

Abrams, Meyer Howard. *The Mirror and the Lamp: Romantic Theory and the Critical Tradition.* London: Oxford University Press, 1971.

———. *Natural Supernaturalism: Tradition and Revolution in Romantic Literature.* New York: W. W. Norton, 1973.

Aguilar, Luis E., ed. *Marxism in Latin America.* Philadelphia: Temple University Press, 1978.

Agustini, Delmira. *Poesías completas.* Edición de Magdalena García Pinto. Madrid: Cátedra, 1993.

Aiken, Henry D. "Philosophy and Ideology in the Nineteenth Century." In *The Age of Ideology: The 19th Century Philosophers.* Selected, with introduction and interpretive commentary, by Henry D. Aiken. New York: New American Library, 1956. 13–26.

Allegra, Giovanni. "El wagnerismo en la exégesis española de primeros de siglo." *Arbor* 116, no. 456 (1983): 79–91.

Alonso, Carlos J. *The Spanish American Regional Novel: Modernity and Autochthony.* Cambridge: Cambridge University Press, 1990.

Anderson Imbert, Enrique. *Historia de la literatura hispanoamericana.* 2nd ed. 2 vols. Mexico: Fondo de Cultura Económica, 1970.

———. *La originalidad de Rubén Darío.* Buenos Aires: Centro Editor de América Latina, 1967.

Bays, Gwendolyn. *The Orphic Vision: Seer Poets from Novalis to Rimbaud.* Lincoln: University of Nebraska Press, 1964.

Bell-Villada, Gene H. *Art for Art's Sake and Literary Life: How Politics and Markets Helped Shape the Ideology and Culture of Aestheticism, 1790–1990.* Lincoln: University of Nebraska Press, 1996.

Benítez, Rubén. "La expresión de lo primitivo en 'Estival', de Darío." *Revista Iberoamericana* 33 (1967): 237–249.

Bennett, John M. "Yo sé de Egipto y Nigricia." In *Antología comentada del modernismo.* Ed. Francisco E. Porrata and Jorge A. Santana. Sacramento: California State University, 1974. 31–36.

Beverley, John, and José Oviedo, eds. *The Postmodernism Debate in Latin America.* Trans. Michael Aronna. Special issue of *Boundary 2: An International Journal of Literature and Culture* 20, no. 3 (1993).

Booker, M. Keith. *Vargas Llosa among the Postmodernists.* Gainesville: University Press of Florida, 1994.

Borges, Jorge Luis. *Leopoldo Lugones.* Buenos Aires: Editorial Pleamar, 1965.

Calinescu, Matei. *Five Faces of Modernity: Modernism, Avant-Garde, Decadence, Kitsch, Postmodernism.* Durham: Duke University Press, 1987.

Cano Gaviria, Ricardo. *José Asunción Silva: Una vida en clave de sombra.* Caracas: Monte Avila, 1992.

Caracciolo Trejo, E. "Lectura de 'Himno a la luna' de Lugones." *Revista Iberoamericana* 44 (1978): 111–117.

Carlos, Alberto J. "'Divagación': La geografía erótica de Rubén Darío." *Revista Iberoamericana* 33 (1967): 293–313.

Carpentier, Alejo. *El acoso.* Madrid: Alfaguara, 1983.

———. *The Chase.* Trans. Alfred MacAdam. New York: Noonday Press, 1990.

———. *Explosion in a Cathedral.* Trans. John Sturruck. New York: Harper and Row, 1979.

———. *Reasons of State.* Trans. Frances Partridge. New York: Knopf, 1976.

———. *El recurso del método.* Mexico City: Siglo Veintiuno Editores, 1974.

———. *El siglo de las luces.* Ed. Ambrosio Fornet. Madrid: Ediciones Cátedra, 1989.

Carranza, María Mercedes. "Silva y el modernismo." In *Obra poética.* By José Asunción Silva. Madrid: Ediciones Hiperión, 1996. 13–25.

Casal, Julián del. *The Poetry of Julián del Casal: A Critical Edition.* Ed. Robert Jay Glickman. 2 vols. Gainesville: University Presses of Florida, 1976–1978.

Castillo, Ana Hernández del. *Keats, Poe, and the Shaping of Cortázar's Mythopoesis.* Purdue University Monographs in Romance Languages. Amsterdam: J. Benjamins, 1981.

Chocano, José Santos. *Obras completas.* Compiladas, anotadas y prologadas por Luis Alberto Sánchez. Madrid: Aguilar, 1954.

Contino, Ferdinand V. "Preciosismo y decadentismo en *De sobremesa* de José Asunción Silva." In *Estudios críticos sobre la prosa modernista hispanoamericana.* Ed. José Olivio Jiménez. New York: Eliseo Torres, 1975. 135–155.

Cortázar, Julio. *Hopscotch.* Trans. Gregory Rabassa. New York: Avon Books, 1975.

———. "Literatura en la revolución y revolución en la literatura: Algunos malentendidos a liquidar." In *Literatura en la revolución y revolución en la literatura.* By Oscar Collazos, Julio Cortázar, and Mario Vargas Llosa. Mexico: Siglo Veintiuno Editores, 1970. 38–77.

———. *Rayuela.* Buenos Aires: Editorial Sudamericana, 1963.

Darío, Rubén. *Azul. Prosas profanas.* Edition, study, and notes by Andrew P. Debicki and Michael J. Doudoroff. Madrid: Editorial Alhambra, 1985.

———. *Obras completas.* Ed. M. Sanmiguel Raimúndez. 5 vols. Madrid: Afrodisio Aguado, 1950.

———. *Poesía.* Ed. Ernesto Mejía Sánchez. Caracas: Biblioteca Ayacucho, 1977.

———. *Selected Poems of Rubén Darío.* Trans. Lysander Kemp. Prologue by Octavio Paz. Austin: University of Texas Press, 1988.

Davison, Ned J. *The Concept of Modernism in Hispanic Criticism.* Boulder, Colo.: Pruett Press, 1966.

De Garganta, Juan. "La política en la poesía de Silva." In *Leyenda a Silva.* Compilación y prólogo de Juan Gustavo Cobo Borda. Bogotá: Instituto Caro Cuervo, 1994. 47–76.

Delgado, Washington. "Situación social de la poesía de Rubén Darío." *Cuadernos Hispanoamericanos* 312 (1976): 575–589.

Díaz-Plaja, Guillermo. *Modernismo frente a noventa y ocho.* Madrid: Espasa-Calpe, 1951.

Dmitrowicz, Gregory. "El concepto de 'espiritualización' en Lugones." *Revista Canadiense de Estudios Hispánicos* 7, no. 3 (1983): 387–392.

Durán, Manuel. *Genio y figura de Amado Nervo.* Buenos Aires: EUDEBA, 1986.

Dyson, John P. "Tragedia dariana: La princesa de la eterna espera." *Atenea* 16, no. 415–416 (1967): 309–319.

Eguren, José María. *Poesías completas.* Lima: Colegio Nacional de Varones "José María Eguren," 1952.

Eire, Ana. "La experiencia de la escritura: El discurso literario en la lectura de Severo Sarduy." Diss. Vanderbilt University, 1992.

———. *Severo Sarduy: Searching for an Art of Living.* Forthcoming.

Ellis, Keith. *Critical Approaches to Rubén Darío.* Toronto: University of Toronto Press, 1974.

Faurie, Marie-Josèphe. *Le Modernisme hispano-américain et ses sources françaises.* Paris: Centre de Recherches de l'Institut Hispanique, 1966.

Fernández-Morera, D. "The Term 'Modernism' in Literary History." In *Proceedings of the Xth Congress of the International Comparative Literature Association.* Ed. Anna Balakian and James J. Wilhelm. New York: Garland, 1985. 271–278.

Fernández Retamar, Roberto. *Introducción a José Martí.* Havana: Casa de las Américas, 1978.

Fiore, Dolores Ackel. *Rubén Darío in Search of Inspiration: Greco-Roman Mythology in His Stories and Poetry.* New York: Las Américas Publishing Co., 1963.

Forcadas, Alberto. "El romancero español, Lope de Vega, Góngora y Quevedo y sus posibles resonancias en 'Sonatina' de Rubén Darío." *Quaderni Ibero-Americani* 41 (1972): 1–6.

———. "El romancero español y posible influjo de algunos clásicos castellanos en 'Sonatina' de Rubén Darío." *Revista de Estudios Hispánicos* 8 (1974): 3–21.

Forster, Merlin H. *Historia de la poesía hispanoamericana.* Clear Creek, Ind.: American Hispanist, 1981.

Foster, David William. "Aeternum vale." In *Antología comentada del modernismo.* Ed. Francisco E. Porrata and Jorge A. Santana. Sacramento: California State University, 1974. 375–381.

Foucault, Michel. *The Order of Things: An Archaeology of the Human Sciences.* New York: Vintage Books, 1973.

Franco, Jean. *César Vallejo: The Dialectics of Poetry and Silence.* Cambridge: Cambridge University Press, 1976.

Fraser, Howard M. "Apocalyptic Vision and Modernism's Dismantling of Scientific Discourse: Lugones's 'Yzur.'" *Hispania* 79 (March 1996): 8–19.

———. *In the Presence of Mystery: Modernist Fiction and the Occult.* North Carolina Studies in the Romance Languages and Literatures, no. 240. Chapel Hill: Department of Romance Languages, University of North Carolina, 1992.

Fuentes, Carlos. *The Campaign.* Trans. Alfred MacAdam. New York: Farrar, Strauss, Giroux, 1991.

———. *La campaña.* Madrid: Narrativa Mondadori, 1990.

———. "Cervantes, or The Critique of Reading." In *Myself with Others: Selected Essays.* New York: Noonday Press, 1988. 49–71.

———. "La literatura es revolucionaria y política en un sentido profundo." *Cuadernos Americanos* 259, no. 2 (1985): 12–16.

———. *La nueva novela hispanoamericana.* Mexico: Cuadernos de Joaquín Mortiz, 1974.

———. Prologue. *Ariel.* By José Enrique Rodó. Trans. Margaret Sayers Peden. Austin: University of Texas Press, 1988. 13–28.

Furness, Raymond. *Wagner and Literature.* New York: St. Martin's Press, 1982.

García Márquez, Gabriel. *The Autumn of the Patriarch.* Trans. Gregory Rabassa. New York: Avon Books, 1976.

———. *Cien años de soledad.* Ed. Jacques Joset. Madrid: Cátedra, 1984.

———. *One Hundred Years of Solitude.* Trans. Gregory Rabassa. New York: Harper and Row, 1970.

———. *El otoño del patriarca.* Barcelona: Plaza y Janes, 1975.

Gay, Peter, ed. Introduction. *The Freud Reader.* By Sigmund Freud. New York: W. W. Norton, 1989. xiii–xxix.

Glickman, Robert Jay. "José Asunción Silva ante los avances tecnológicos de su época." *Canadian Journal of Latin American Studies* 1, nos. 1–2 (1976): 180–190.

———. "Neurosis." *Antología comentada del modernismo.* Ed. Francisco E. Porrata and Jorge A. Santana. Sacramento: California State University, 1974. 171–175.

Godzich, Wlad. "Afterword: Reading against Literacy." In *The Postmodern Explained.* By Jean-François Lyotard. Translations edited by Julian Pefanis and Morgan Thomas. Minneapolis: University of Minnesota Press, 1992. 109–136.

González, Aníbal. *La crónica modernista hispanoamericana.* Madrid: José Porrúa Turanzas, 1983.

———. *La novela modernista hispanoamericana.* Madrid: Editorial Gredos, 1987.

González Echevarría, Roberto. *Alejo Carpentier: The Pilgrim at Home.* Ithaca, N.Y.: Cornell University Press, 1977.

———. "Martí y su 'Amor de ciudad grande': Notas hacia la poética de *Versos libres.*" In *Isla a su vuelo fugitivo: Ensayos críticos sobre literatura hispanoamericana.* Madrid: José Porrúa Turanzas, 1983. 27–42.

———. "Memoria de apariencias y ensayo de *Cobra.*" *Relecturas: Estudios de literatura cubana.* Caracas: Monte Avila, 1976. 129–152.

———. "Modernidad, modernismo y nueva narrativa: *El recurso del método.*" *Revista Interamericana de Bibliografía / Interamerican Review of Bibliography* 30 (1980): 157–163.

———. "La nación desde *De donde son los cantantes* a *Pájaros de la playa.*" *Cuadernos Hispanoamericanos* 563 (1997): 55–67.

———. Preamble. *The Voice of the Masters: Writing and Authority in Modern Latin American Literature.* Austin: University of Texas Press, 1985. 1–7.

———. "El primer relato de Severo Sarduy." In *Isla a su vuelo fugitivo: Ensayos críticos sobre literatura hispanoamericana.* Madrid: José Porrúa Turanzas, 1983. 123–143.

———. "*Los reyes*: Cortázar's Mythology of Writing." In *The Voice of the Masters: Writing and Authority in Modern Latin American Literature.* Austin: University of Texas Press, 1985. 98–109.

———. *La ruta de Severo Sarduy.* Hanover, N.H.: Ediciones del Norte, 1987.

González Martínez, Enrique. *La apacible locura.* In *Memorias y autobiografías de escritores mexicanos.* Ed. Raymundo Ramos. Mexico: Universidad Nacional Autónoma de México, 1967. 151–168.

———. *Obras completas.* Edición, prólogo y notas de Antonio Castro Leal. Mexico: Colegio Nacional, 1971.

González-Rodas, Publio. "Presencia de Sarmiento en Rubén Darío." *Revista Iberoamericana* 38 (1972): 287–299.

González Stephan, Beatriz. "Fundar el estado / narrar la nación (*Venezuela heroica* de Eduardo Blanco)." *Revista Iberoamericana* 63 (1997): 33–46.

Gullón, Ricardo. "Exotismo y modernismo." In *Estudios críticos sobre el modernismo.* Ed. Homero Castillo. Madrid: Gredos, 1974. 279–298.

Gutiérrez Giradot, Rafael. *Modernismo.* Barcelona: Montesinos, 1983.

Gutiérrez Mouat, Ricardo. "La presencia de ciertos textos de Darío en *Residencia en la tierra.*" *Hispamérica* 13, no. 39 (1984): 85–93.

Gutiérrez Nájera, Manuel. *Obras: Crítica literaria.* Ed. Ernesto Mejía Sánchez. Mexico: Universidad Nacional Autónoma de México, 1959.

———. *Poesías.* Prologue by Justo Sierra. 2 vols. Mexico: Librería de la Vda. de Ch. Bouret, n.d.

Hatzfeld, Helmut A. "Rubén Darío (1867–1916): 'Sonatina.'" In *Explicación de textos literarios.* Sacramento: California State University, 1973. 155–162.

Hayles, N. Katherine. *Chaos Bound: Orderly Disorder in Contemporary Literature and Science.* Ithaca, N.Y.: Cornell University Press, 1990.

———. *The Cosmic Web: Scientific Field Models and Literary Strategies in the Twentieth Century.* Ithaca, N.Y.: Cornell University Press, 1984.

———, ed. *Chaos and Order: Complex Dynamics in Literature and Science.* Chicago: University of Chicago Press, 1991.

Henríquez Ureña, Max. *Breve historia del modernismo.* Mexico: Fondo de Cultura Económica, 1954.

Hernández del Castillo, Ana. *Keats, Poe, and the Shaping of Cortázar's Mythopoesis.* Purdue University Monographs in Romance Languages. Amsterdam: John Benjamins, 1981.

Herrera y Reissig, Julio. *Poesía completa y prosa selecta.* Prólogo de Idea Vilariño; edición, notas y cronología de Alicia Migdal. Caracas: Biblioteca Ayacucho, 1978.

Herrero, Javier. "Fin de siglo y modernismo. La virgen y la hetaira." *Revista Iberoamericana* 46 (1980): 29–50.

Hewitt, Sandra, and Nancy Abraham Hall. "Leopoldo Lugones and H. P. Blavatsky: Theosophy in the 'Ensayo de una cosmogonía en diez lecciones.'" *Revista de Estudios Hispánicos* 18, no. 3 (1984): 335–343.

Higgins, James. "The Rupture between Poet and Society in the Work of José María Eguren." *Kentucky Romance Quarterly* 20 (1973): 59–74.

Ingwersen, Sonya A. *Light and Longing: Silva and Darío, Modernism and Religious Heterodoxy.* New York: Peter Lang, 1986.

Jaimes Freyre, Mireya. "Universalismo y romanticismo en un poeta 'modernista': Ricardo Jaimes Freyre." *Revista Hispánica Moderna* 31 (1965): 236–246.

Jaimes Freyre, Ricardo. *Poesías completas.* La Paz: Ministerio de Educación y Bellas Artes, 1957.

Jameson, Fredric. "Postmodernism and Consumer Society." In *The Anti-Aesthetic: Essays on Postmodern Culture.* Ed. Hal Foster. Port Townsend, Wash.: Bay Press, 1983. 111–126.

———. *Postmodernism, or The Cultural Logic of Late Capitalism.* Durham: Duke University Press, 1991.

Jensen, Theodore W. "*Modernista* Pythagorean Literature: The Symbolist Inspiration." Ed. Roland Grass and William R. Risley. In *Waiting for Pegasus: Studies of the Presence of Symbolism and Decadence in Hispanic Letters.* Macomb: Western Illinois University, 1979. 169–179.

Jitrik, Noé. *Las contradicciones del modernismo: Productividad poética y situación sociológica.* Mexico: Colegio de México: 1978.

Jrade, Cathy L. "El Modernismo y la Generación del '98: Ideas afines, creencias divergentes." *Texto Crítico* 14, no. 38 (1988): 15–29.

———. "Modernist Poetry." In *Cambridge History of Latin American Literature.* Ed. Roberto González Echevarría and Enrique Pupo-Walker. Vol. 2. Cambridge: Cambridge University Press, 1996. 7–68.

———. *Rubén Darío and the Romantic Search for Unity: The Modernist Recourse to Esoteric*

Tradition. Austin: University of Texas Press, 1983. Augmented Spanish version, *Rubén Darío y la búsqueda romántica de la unidad: El recurso modernista a la tradición esotérica.* Mexico: Fondo de Cultura Económica, 1986.

———. "El significado de un vínculo textual inesperado: *Rayuela* y 'Tuércele el cuello al cisne.'" *Revista Iberoamericana* 47 (July–Dec. 1981): 145–154.

———. "Socio-Political Concerns in the Poetry of Rubén Darío." *Latin American Literary Review* 36 (1990): 36–49.

Kaliman, Ricardo J. "La carne y el mármol: Parnaso y simbolismo en la poética modernista hispanoamericana." *Revista Iberoamericana* 55 (1989): 17–32.

Kirkpatrick, Gwen. "Art and Politics in Lugones' Early Journalism." *Discurso Literario: Revista de Temas Hispánicos* 3, no. 1 (1985): 81–95.

———. *The Dissonant Legacy of Modernismo: Lugones, Herrera y Reissig, and the Voices of Modern Spanish American Poetry.* Berkeley: University of California Press, 1989.

Kristal, Efraín. "Dialogues and Polemics: Sarmiento, Lastarria, and Bello." In *Sarmiento and His Argentina.* Ed. Joseph T. Criscenti. Boulder, Colo.: Lynne Rienner Publishers, 1993. 61–70.

Large, David C., and William Weber, eds. *Wagnerism in European Culture and Politics.* Ithaca, N.Y.: Cornell University Press, 1984.

Leal, Luis. "Situación de Amado Nervo." *Revista Iberoamericana* 36 (1970): 485–494.

Loveluck, Juan. "*De sobremesa,* novela desconocida del modernismo." *Revista Iberoamericana* 31 (1965): 17–32.

Lugones, Leopoldo. *Obras poéticas completas.* Madrid: Aguilar, 1959.

Lyotard, Jean-François. *The Postmodern Condition: A Report on Knowledge.* Trans. Geoff Bennington and Brian Massumi. Foreword by Frederic Jameson. Minneapolis: University of Minnesota Press, 1984.

———. *The Postmodern Explained: Correspondence, 1982–1985.* Minneapolis: University of Minnesota Press, 1992.

Mapes, Erwin K. *L'Influence française dans l'oeuvre de Rubén Darío.* Paris: Champion, 1925.

Marasso, Arturo. *Rubén Darío y su creación poética.* Buenos Aires: Kapeluz, 1954.

Mariátegui, José Carlos. "El proceso de la literatura." In *Siete ensayos de interpretación de la realidad peruana.* Havana: Casa de las Américas, 1975. 207–325.

Marini Palmieri, Enrique. "Esoterismo en la obra de Leopoldo Lugones." *Cuadernos Hispanoamericanos* 458 (1988): 79–95.

Martí, José. *Ismaelillo; Versos libres; Versos sencillos.* Edición de Ivan A. Schulman. Madrid: Ediciones Cátedra, 1992.

———. *José Martí, Major Poems: A Bilingual Edition.* English translation by Elinor Randall. Edited, with an introduction, by Philip S. Foner. New York: Holmes and Meier Publishers, 1982.

———. *Obras completas.* 27 vols. Havana: Editorial Nacional de Cuba, 1963–1973.

———. "Prólogo para *El poema del Niágara.*" In *Nuestra América.* Introd. Pedro Henríquez Ureña. Buenos Aires: Losada, 1980. 103–125.

Marún, Gioconda. "*De sobremesa*: El vértigo de lo invisible." *Thesaurus* 140, no. 2 (1985): 361–374.

McGowan, John. *Postmodernism and Its Critics.* Ithaca, N.Y.: Cornell University Press, 1991.

Meyer-Minnemann, Klaus. "*De sobremesa,* de José Asunción Silva." In *La novela hispanoamericana de fin de siglo.* Mexico: Fondo de Cultura Económica, 1991. 40–73.

Miller, Beth, ed. *Women in Hispanic Literature: Icons and Fallen Idols.* Berkeley: University of California Press, 1983.

Miller, Roger W. "The Evolution of Parnassianism in Two Major Works of Rubén Darío, *Azul* and *Prosas profanas.*" Diss. University of Colorado, 1971.

Molloy, Sylvia. "Dos lecturas del cisne: Rubén Darío y Delmira Agustini." In *La sartén por el mango: Encuentro de escritoras latinoamericanas.* Ed. Patricia Elena González and Eliana Ortega. Río Piedras, Puerto Rico: Ediciones Huracán, 1984. 57–69.

Monguió, Luis. "De la problemática del modernismo: La crítica y el 'cosmopolitismo'." *Revista Iberoamericana* 28 (1962): 75–86.

———. *La poesía postmodernista peruana.* Berkeley: University of California Press, 1954.

Moretić, Yerko. "Acerca de las raíces ideológicas del modernismo hispanoamericano." *Philologica Pragensia* 8 (1965): 45–53.

Munárriz, Jesús. "De esta edición y sus características." In *Obra poética.* By José Asunción Silva. Madrid: Ediciones Hiperión, 1996. 293–300.

Natella, Arthur. "Towards a Definition of Latin American Modernism—Debate and Reconciliation." *Romanticism Past and Present* 8, no. 2 (1984): 23–38.

Navarro Tomás, Tomás. "Análisis de la Sonatina." In *Estudios de fonología española.* New York: Las Américas Publishing Co., 1966. 192–202.

Neruda, Pablo. *Canto general.* Ed. Enrico Mario Santí. Madrid: Ediciones Cátedra, 1992.

———. *Canto General.* Trans. by Jack Schmitt. Berkeley: University of California Press, 1991.

Nervo, Amado. *Obras completas.* Ed. Francisco González Guerrero y Alfonso Méndez Plancarte. 2 vols. Madrid: Aguilar, 1956.

Nietzsche, Friedrich Wilhelm. *The Portable Nietzsche.* Selected and translated, with an introduction, prefaces, and notes, by Walter Kaufmann. New York: Penguin Books, 1954.

Núñez, Estuardo. *José María Eguren: Vida y obra.* Lima: P. L. Villanueva, 1964.

O'Brien, Pat. "'Sonatina'—Manifesto of Modernism." *South Central Bulletin* 42 (1982): 134–136.

O'Hara, Edgar. "*De sobremesa,* una divagación narrativa." *Revista Chilena de Literatura* 27–28 (Apr.–Nov. 1986): 221–227.

Orjuela, Héctor. Cronología. In *Obra poética.* By José Asunción Silva. Madrid: Ediciones Hiperión, 1996.

Ortega, Julio. "José María Eguren." *Cuadernos Hispanoamericanos* 247 (1970): 60–85.

Pacheco, José Emilio, ed. *Antología del modernismo, 1884–1921.* 2 vols. Mexico: Universidad Nacional Autónoma de México, 1978.

Palau de Nemes, Graciela. "Tres momentos del neomisticismo poético del 'siglo modernista': Darío, Jiménez y Paz." In *Estudios sobre Rubén Darío*. Ed. Ernesto Mejía Sánchez. Mexico: Fondo de Cultura Económica, 1968. 536–552.

Paraíso de Leal, Isabel. "Teoría y práctica del verso libre en Ricardo Jaimes Freyre." *Revista Española de Lingüística* 12, no. 2 (1982): 311–319.

Paz, Octavio. "El caracol y la sirena." In *Cuadrivio*. Mexico: Joaquín Mortiz, 1965. 11–65.

———. *Children of the Mire: Modern Poetry from Romanticism to the Avant-Garde*. Trans. Rachel Phillips. Cambridge: Harvard University Press, 1974.

———. *Los hijos del limo: Del romanticismo a la vanguardia*. Barcelona: Seix Barral, 1974.

———. "Literatura de fundación." In *Puertas al campo*. Mexico: Universidad Nacional Autónoma de México, 1966. 11–19.

———. "The Siren and the Seashell." In *"The Siren and the Seashell" and Other Essays on Poets and Poetry*. Trans. Lysander Kemp and Margaret Sayers Peden. Austin: University of Texas Press, 1976. 17–56.

Pérus, Françoise. *Literatura y sociedad en América Latina: El modernismo*. Havana: Casa de las Américas, 1976.

Petit, Anne. "Mito y tiempo en tres obras posmodernas de Carlos Fuentes." Diss. Vanderbilt University, 1998.

Phillips, Allen W. *Alejandro Sawa: Mito y realidad*. Madrid: Ediciones Turner, 1976.

———. "Cuatro poetas hispanoamericanos entre el modernismo y la vanguardia." *Revista Iberoamericana* 55, nos. 146–147 (1989): 427–449.

———. "Dos protagonistas: Un poeta y un escultor." *Temas del modernismo hispánico y otros estudios*. Madrid: Gredos, 1974. 264–271.

Picon Garfield, Evelyn. "*De sobremesa*: José Asunción Silva: El diario íntimo y la mujer prerrafaelita." *Nuevos asedios al modernismo*. Ed. Iván A. Schulman. Madrid: Taurus Ediciones, 1987. 262–281.

Picon Garfield, Evelyn, and Iván A. Schulman. *"Las entrañas del vacío": Ensayos sobre la modernidad hispanoamericana*. Mexico: Cuadernos Americanos, 1984.

———. "José Martí: El *Ismaelillo* y las prefiguraciones vanguardistas del modernismo." In *"Las entrañas del vacío": Ensayos sobre la modernidad hispanoamericana*. Mexico: Cuadernos Americanos, 1984. 79–96.

Plotnitsky, Arkady. *Complementarity: Anti-Epistemology after Bohr and Derrida*. Durham: Duke University Press, 1994.

———. *In the Shadow of Hegel: Complementarity, History, and the Unconscious*. Gainesville: University Press of Florida, 1993.

———. *Reconfigurations: Critical Theory and General Economy*. Gainesville: University Press of Florida, 1993.

Plotnitsky, Arkady, and Barbara H. Smith, eds. *Mathematics, Science, and Postclassical Theory*. Durham: Duke University Press, 1995.

Podesta, Bruno. "Hacia una conceptualización ideológica del modernismo hispánico." *Cuadernos Americanos* 195 (1974): 227–237.

Poirer, Richard. "The Difficulties of Modernism and the Modernism of Difficulty." In *Images and Ideas in American Culture*. Ed. Arthur Edelstein. Hanover, N.H.: Brandeis University Press, 1979. 124–140.

Predmore, Michael P. "A Stylistic Analysis of 'Lo fatal.'" *Hispanic Review* 39 (1971): 433–438.

Rama, Angel. *La ciudad letrada*. Hanover, N.H.: Ediciones del Norte, 1984.

———. *Las máscaras democráticas del modernismo*. Montevideo: Fundación Angel Rama, 1985.

———. *Los poetas modernistas en el mercado económico*. Montevideo: Universidad de la República, 1968.

———. *Rubén Darío y el modernismo (circunstancia socioeconómica de un arte americano)*. Caracas: Ediciones de la Biblioteca de la Universidad Central de Venezuela, 1970.

Ramos, Julio. *Desencuentros de la modernidad en América Latina: Literatura y política en el siglo XIX*. Mexico: Fondo de Cultura Económica, 1989.

Raymond, Marcel. *From Baudelaire to Surrealism*. London: Methuen, 1970.

Real de Azúa, Carlos. "El modernismo literario y las ideologías." *Escritura: Teoría y Crítica Literarias* 3 (1977): 41–75.

Rivera Meléndez, Blanca Margarita. "Poetry and the Machinery of Illusion: José Martí and the Poetics of Modernity." Diss. Cornell University, 1990.

Rivero, Eliana S. "La duquesa Job." In *Antología comentada del modernismo*. Ed. Francisco E. Porrata and Jorge A. Santana. Sacramento: California State University, 1974. 92–99.

Rodó, José Enrique. *Ariel*. Mexico: Editorial Porrúa, 1968.

———. *Ariel*. Trans. Margaret Sayers Peden. Foreword by James W. Symington. Prologue by Carlos Fuentes. Austin: University of Texas Press, 1988.

Rodríguez Monegal, Emir. *El otro Andrés Bello*. Caracas: Monte Avila Editores, 1969.

Rodríguez-Peralta, Phyllis. "The Modernism of José María Eguren." *Hispania* 56 (1973): 222–229.

Rodríguez-Sardiñas, Orlando. "Delectación morosa." In *Antología comentada del modernismo*. Ed. Francisco E. Porrata and Jorge A. Santana. Sacramento: California State University, 1974. 457–460.

Roggiano, Alfredo A. "Modernismo: Origen de la palabra y evolución de un concepto." In *In Honor of Boyd G. Carter*. Ed. Catherine Vera and George R. McMurray. Laramie: University of Wyoming, 1981. 93–103.

Rojas, Víctor J. "Holocausto." In *Antología comentada del modernismo*. Ed. Francisco E. Porrata and Jorge A. Santana. Sacramento: California State University, 1974. 460–464.

Salinas, Pedro. *La poesía de Rubén Darío*. Buenos Aires: Editorial Losada, 1948.

———. "El problema del modernismo en España, o un conflicto entre dos espíritus." *Literatura española siglo XX*. Madrid: Alianza, 1970. 13–25.

Sánchez-Castañer, Francisco. *Rubén Darío y el mar*. Alicante: Cátedra "Mediterráneo," 1969.

Sanín Cano, Baldomero. *El oficio del lector*. Caracas: Biblioteca Ayacucho, n.d.

Santí, Enrico Mario. "*Ismaelillo*, Martí y el modernismo." *Revista Iberoamericana* 52 (1986): 811–840.

Santos Molano, Enrique. *El corazón del poeta*. Bogotá: Nuevo Rumbo Editores, 1992.

Sarduy, Severo. *De donde son los cantantes.* Ed. Roberto González Echevarría. Madrid: Cátedra, 1993.

———. *From Cuba with a Song.* Trans. Suzanne Jill Levine. Los Angeles: Sun and Moon Press, 1994.

Sarduy, Severo, Tomás Segovia, and Emir Rodríguez Monegal. "Nuestro Rubén Darío." *Mundo Nuevo* no. 7 (1967): 33–46.

Sarrailh, Michèle. "Apuntes sobre el mito dariano en *El otoño del patriarca.*" In *Actas del simposio internacional de estudios hispánicos: Budapest, 18–19 de agosto de 1976.* Ed. Mátyás Horányi. Budapest: Editorial de la Academia de Ciencias de Hungría, 1978. 435–458.

———. "Rubén Darío y el modernismo en *La increíble y triste historia de la cándida Eréndira.*" In *XVII Congreso del Instituto Internacional de Literatura Iberoamericana: Sesión de Madrid.* Vol. 1. Madrid: Ediciones Cultura Hispánica del Centro Iberoamericano de Cooperación, 1978. 707–724.

Saurat, Denis. *Literature and Occult Tradition: Studies in Philosophical Poetry.* 1930. Port Washington, N.Y.: Kennikat Press, 1966.

Schulman, Iván A. *Génesis del modernismo: Martí, Nájera, Silva, Casal.* Mexico: Colegio de México, 1966.

———. "Modernismo/modernidad: Metamorfosis de un concepto." In *Nuevos asedios al modernismo.* Ed. Iván A. Schulman. Madrid: Taurus, 1987. 11–38.

———. *Relecturas martianas: Narración y nación.* Atlanta: Rodopi, 1994.

Schulman, Iván A., and Manuel Pedro González. *Martí, Darío y el modernismo.* Prologue by Cintio Vitier. Madrid: Gredos, 1969.

Senior, John. *The Way Down and Out: The Occult in Symbolist Literature.* Ithaca, N.Y.: Cornell University Press, 1959.

Shroder, Maurice Z. *Icarus: The Image of the Artist in French Romanticism.* Cambridge: Harvard University Press, 1961.

Silva, José Asunción. *De sobremesa.* Prólogo de Gabriel García Márquez. Madrid: Ediciones Hiperión, 1996.

———. *Obra poética.* Testimonio de Alvaro Mutis. Introducción de María Mercedes Carranza. Cronología de Héctor H. Orjuela. Edición de Jesús Munárriz. Madrid: Ediciones Hiperión, 1996.

Skyrme, Raymond. "Darío's *Azul* . . . : A Note on the Derivation of the Title." *Romance Notes* 10 (1969): 73–66.

———. *Rubén Darío and the Pythagorean Tradition.* Gainesville: University Presses of Florida, 1975.

Sommer, Doris. *Foundational Fictions: The National Romances of Latin America.* Berkeley: University of California Press, 1991.

Stephens, Doris. *Delmira Agustini and the Quest for Transcendence.* Montevideo: Ediciones Géminis, 1975.

Trigo, Benigno. "La función crítica del discurso alienista en *De sobremesa* de José Asunción Silva." *Hispanic Journal* 15, no. 1 (1994): 133–146.

Trueblood, Alan S. "Rubén Darío: The Sea and the Jungle." *Comparative Literature Studies* 4 (1967): 425–456.

———. "Wilde y Valencia: *La balada de la Cárcel de Reading*." In *Estudios: Edición en Homenaje a Guillermo Valencia, 1873–1973*. Ed. Hernán Torres. Cali, Colombia: Carvajal y Compañía, 1976. 141–191.

Valencia, Guillermo. *Obras poéticas completas*. Madrid: Aguilar, 1955.

Valera, Juan. Carta-Prólogo. *Azul* . . . . By Rubén Darío. Madrid: Espasa-Calpe, 1937. 9–25.

Vélez, Joseph F. "Vieja llave." In *Antología comentada del modernismo*. Ed. Francisco E. Porrata and Jorge A. Santana. Sacramento: California State University, 1974. 392–397.

Villanueva-Collado, Alfredo. "*De sobremesa* de José Asunción Silva y las doctrinas esotéricas en la Francia de fin de siglo." *Revista de Estudios Hispánicos* 221, no. 2 (1987): 9–22.

———. "La funesta Helena: Intertextualidad y caracterización en *De sobremesa*, de José Asunción Silva." *Explicación de textos literarios* 22, no. 1 (1993–1994): 63–71.

Vitier, Cintio. "En la mina martiana" (prologue). In *Martí, Darío y el modernismo*. By Iván A. Schulman and Manuel Pedro González. Madrid: Gredos, 1969. 9–21.

———. *Lo cubano en la poesía*. Havana: Instituto del Libro, 1970.

———. "Vallejo y Martí." *Revista de Crítica Literaria Latinoamericana* 7, no. 13 (1981): 95–98.

Yurkievich, Saúl. *Celebración del modernismo*. Barcelona: Tusquets, 1976.

———. "El efecto manifestario, una clave de modernidad." In *Recreaciones: Ensayos sobre la obra de Rubén Darío*. Prólogo y edición de Iván A. Schulman. Hanover, N.H.: Ediciones del Norte, 1992. 213–228.

———. "Rubén Darío, precursor de la vanguardia." *Literatura de la emancipación hispanoamericana y otros ensayos*. Lima: Universidad Nacional Mayor de San Marcos, Dirección Universitaria de Biblioteca y Publicaciones, 1972. 117–131.

———. "Rubén Darío y la modernidad." *Plural* 9 (1972): 37–41.

Zamora Vicente, Alonso. "'Divagación': Aclaración sobre el modernismo." In *El comentario de textos*. By Emilio Alarcos et al. Madrid: Castalia, 1973. 167–193.

Zavala, Iris. *Colonialism and Culture: Hispanic Modernisms and the Social Imaginary*. Bloomington: Indiana University Press, 1992.

———. "Estudio preliminar." In *Iluminaciones en la sombra*. By Alejandro Sawa. Madrid: Editorial Alhambra, 1977. 3–66.

———. "1898, Modernismo and the Latin American Revolution." *Revista Chicano-Riqueña* 3 (1975): 43–47.

Index